PERGAMON INTERNATIONAL LIBRARY
of Science, Technology, Engineering and Social Studies
The 1000-volume original paperback library in aid of education, industrial training and the enjoyment of leisure
Publisher: Robert Maxwell, M.C.

INTERNATIONAL SERIES IN
NATURAL PHILOSOPHY
General Editor: D TER HAAR

VOLUME 97

STELLAR FORMATION

Other titles of interest

1. MEADOWS: Stellar Evolution 2nd Edition
2. PACHOLCZYK: Radio Galaxies
3. HEY: Radio Universe 2nd Edition
4. CLARK AND STEVENSON: Historical Supernovae
5. SAHADE AND WOOD: Interacting Binary Stars

STELLAR FORMATION

BY

V C REDDISH

Astronomer Royal for Scotland

Director, Royal Observatory, Edinburgh

PERGAMON PRESS

OXFORD · NEW YORK · TORONTO · SYDNEY · PARIS · FRANKFURT

U.K.	Pergamon Press Ltd., Headington Hill Hall, Oxford OX3 0BW, England
U.S.A.	Pergamon Press Inc., Maxwell House, Fairview Park, Elmsford, New York 10523, U.S.A.
CANADA	Pergamon of Canada Ltd., 75 The East Mall, Toronto, Ontario, Canada
AUSTRALIA	Pergamon Press (Aust.) Pty. Ltd., 19a Boundary Street, Rushcutters Bay, N.S.W. 2011, Australia
FRANCE	Pergamon Press SARL, 24 rue des Ecoles, 75240 Paris, Cedex 05, France
FEDERAL REPUBLIC OF GERMANY	Pergamon Press GmbH, 6242 Kronberg-Taunus, Pferdstrasse 1, Federal Republic of Germany

First edition 1978

British Library Cataloguing in Publication Data

Reddish, Vincent Cartledge
Stellar formation.
1. Stars evolution
I. Title
523.8 QB806 78-40550

ISBN 0-08-018062-0 Hardcover
ISBN 0-08-023053-9 Flexicover

In order to make this volume available as economically and as rapidly as possible the author's typescript has been reproduced in its original form. This method unfortunately has its typographical limitations but it is hoped that they in no way distráct the reader.

Printed in Great Britain by William Clowes & Sons Limited London, Beccles and Colchester

CONTENTS

PART 1: DATA

PART 2: THEORY

AUTHOR'S NOTE

When the idea of this book was first mooted in 1966 I mentioned it to a colleague and he said that it was impossible - we didn't know enough about star formation to write a book about it. He was right, but much has happened in the meantime. The attempt stimulated much of my own research and some of that of my colleagues in the years that followed; the rapid development of radio wavelength measurements of emission by interstellar molecules produced a growing body of data in the literature about the cool clouds in regions where star formation appeared to be taking place; and the subject rapidly grew in popularity so that the literature now abounds with relevant observational and theoretical papers. This book covers publications up to the end of 1974.

Any book such as this is a cooperative effort. I benefited from early discussions with J.G. Ireland. Consideration with F. Hoyle and N.C. Wickramasinghe of the possibility that hydrogen might freeze on the surfaces of grains led me to consider the role which grains may play in the processes of star formation and provided the stimulus for the experiments by T.J. Lee which led to much more accurate values for the adsorption energies of hydrogen on cold surfaces. I have had many valuable discussions with colleagues at the Royal Observatory Edinburgh. I am indebted to Dr. D. ter Haar for reading the manuscript with such care and making many helpful suggestions for improvements; to the publishers for their patience; and to Mrs. Jo Barrow who produced the first draft, the second draft, and the manuscript from which this book is reproduced, with extraordinary care and patience.

V.C. Reddish
30 September 1975

"If the end of wisdom is to add
star to star our foolishness is pleasing."

Seven Pillars of Wisdom
T.E. Lawrence

PART 1: DATA

Chapter 1.

INTRODUCTION.

The sense of excitement which comes from search and discovery is infectious. Few of us fail to catch the enthusiasm of those who search for the origin of homo sapiens, the origin of life, the origin of matter; although the relevant disciplines are different, the strong element of curiosity is common and we share an interest in the outcome of the search. Those three topics are of basic and general interest. Chronologically, star formation necessarily comes in between the formations of matter and of life; as do the formations of many other cosmic objects - galaxies, planets, dust grains, molecules. At first sight none appears to hold a special place or to play a key role. The view appears to be widespread amongst astronomers that many features of these and of other phenomena - the difference between elliptical and spiral galaxies, the rate of star formation and the frequency distribution of stellar masses, the chemical compositions of the stars and the gas and the grains, quasars, the cosmic background temperature, cosmic expansion, for examples - are related to each other only in a statistical sense rather than as the result of some deep physical interdependence. Both viewpoints are represented in this book which is intended to survey and to gather together present knowledge on the subject, but it will become apparent to the reader that the author inclines strongly to the view that the statistical relationships not only have a physical origin but that they represent in total a picture of an evolving but stable and self regulating system of galaxies of stars in which the process of star formation itself is a vital element.

The process of forming a star may be said to have ended when a mass of gas has reached a state in which, excepting unusual accidents, it will inevitably become a star of essentially that same mass. Thereafter, it becomes the concern of stellar evolution rather than of star formation.

It is more difficult to define a starting point for the formation of a star. The formation of galaxies not being a part of our present study, it is reasonable to use as a beginning the existence of a cloud of interstellar matter of galactic mass; although it must then be admitted that we immediately face a difficulty in that we are uncertain of the initial values to give to various parameters of the cloud - chemical composition, temperature, density; we must therefore constantly be concerned to discover if the route being pursued depends sensitively on the starting point.

Within these rather ill-defined boundaries we search for answers to several questions.

How and where do stars form; at what rate and with what properties? And more fundamentally - in seeking to determine the conditions necessary for star formation - why do stars form? Is the formation of a star an accident and does the probability of that accident, represented by the rate of star formation, have any particular value; if it does, is that value the result of a chance combination of circumstances peculiar to the way the Universe has developed or has it always been inherent in the properties of the Universe?

Or is star formation an integral part of the physical properties of matter, necessarily having the properties it has as a consequence of the properties possessed by matter itself?

The considerations presented in this book lead the author to conclude that the latter view is correct.

The various theories from various authors gathered together in later chapters cover a range that extends from the entirely statistical to the mainly physical. It is to be doubted if any of them find general acceptance at present, even those concerned with such restricted parts of the process of star formation as, say, the frequency distribution of stellar masses. Nonetheless they may contain among them much of the truth or guidelines to it. They therefore represent what is on offer, and in our present state of understanding of star formation the reader must be left to make his own choice, a choice that may perhaps be changed and guided and finally directed towards the correct solutions, if indeed they occur in whole or in part between these covers, by the increasingly rapid accumulation of relevant data.

Advice is not absent from this book; although it is not restricted to the author's own theories understandably it is biased towards them. These rest on the basis that interstellar grains play a determinate role in the process of star formation. Before mentioning some of the reasons for this it is of interest to consider the present state of knowledge of star formation in comparison to, say, the theories of stellar structure and evolution.

We are approximately at the point corresponding to that reached in these latter subjects half a century ago; then there were suggestions - but no generally accepted view - of the origin of stellar thermal energy; it was argued whether relationships such as the stellar mass-luminosity 'law' represented basic physical properties of the stars or mere chance statistical dependences. The conflict between the equilibrium theories of Eddington, in which negative feedback played such a basic role in determining the properties of stars, and other theories based on the cooling of incompressible fluids, is well known.

In the case of star formation a parallel may be drawn between attempts to explain the frequency distribution of stellar masses and the attempts nearly half a century earlier to account for the stellar mass-luminosity law. It turned out of course that Eddington was right in that a given amount of mass of given composition gathers itself together in such a manner that it emits energy at a given rate, and it draws on its available energy supply as necessary to meet demand, negative feedback ensuring stable equilibrium. The view that the mass-luminosity relation was a statistical rather than a physical relationship was shown to be wrong. Similarly in the case of the frequency distribution of stellar masses, attempts at explanation range from purely statistical to essentially physical and the correct answer has yet to be proved. The present writer considers that the physical explanation is likely to be the correct one - that the frequency distribution of the masses of stars is as much a consequence of the physical process of forming stars as the luminosity of a star is a consequence of a given mass and composition. This view is held of other major properties such as the rate of star formation and the physical and chemical properties of the matter from which stars form, and indeed of the whole process of the physical evolution of a galaxy of stars.

As mentioned above, the theories which these views represent rest on the basis that the interstellar solid grains play a determinate role - it is argued that star formation results from pressure variations brought about by the combination of atoms of interstellar hydrogen gas into molecules or solids on the surfaces of grains, a process critically dependent for its rate on the grain temperature; and that the interaction of the grains with the ambient radiation field determines the processes of the formation of stars and, by introducing several negative feedback loops, controls the values of the parameters associated with them.

These theories of star formation have reached a stage where, given a cosmic volume of space expanding at the observed rate, and containing galactic masses of atomic hydrogen gas with a low grain abundance in volumes of approximately galactic size, the resulting evolution can be predicted using only data of laboratory physics. The predictions are included in the text with those of other relevant hypotheses and theories in chapters 11 to 13 and summarised in chapter 13. They include predictions of the frequency distribution of stellar masses; the rate of star formation; the occurrence of super-massive objects; the growth with time of the helium, heavier element and grain abundances in stars and in the interstellar gas; relationships between the luminosities per unit mass, gas contents, ages and integrated spectra of galaxies; the origin of the difference between elliptical and spiral galaxies; and the average density of galactic matter in the Universe. All the predictions are in good agreement with the observational data; but some of these data are meagre or confused, and the author makes no claim for certainty in the details of his proposals. One matter ought, however, to be emphasised. The theories depend critically on only one factor - the argument that the interstellar grains play a decisive part in star formation through the combination of hydrogen on their surfaces. Exactly how this factor operates will affect the details of the predictions but not their general features.

The contraction of a mass of gas - part of a cloud - to form a star requires the action of pressure variations within the cloud to initiate the condensation; in one way or another this is a common feature of all theories of star formation.

The only realistic alternative to pressure variations induced by combination of hydrogen is turbulence, and heretofore this has been the most widespread and perhaps the most widely accepted basis of theories of star formation. However, the accumulated observations indicate that stars form in dark clouds within which, prior to star formation, turbulence is very low, and no stars have been observed to have formed in the very turbulent regions of clouds such as are generated by expanding HII regions or supernovae or intercloud collisions. Furthermore, calculations indicate that the pressure variations brought about by combination of hydrogen will be larger than those due to turbulence and will operate in the desired sense so that locally the pressure falls as the mass density increases. Consequently, turbulence now appears to be a less likely basis for star formation than the combination of hydrogen into molecules or solids on grain surfaces.

What is the evidence that the grains play this decisive role?

There is nothing conclusive; but there are strong indications.

It is well-known that there is a close correlation between the locations of young stars and dense clouds of dust, correlations that are particularly marked for T Tauri stars, which are believed to be young dwarf stars, and for compact HII regions, which mark the places of recent formation of massive stars. These correlations of course do not show that the presence of dust is necessary for star formation - only that it is present in large quantities and at high space densities where star formation has recently taken place. To prove by its presence that dust is necessary, it has to be shown that star formation will not take place in its absence from clouds of gas of similar total density, and no dense dust-free clouds are known; the conclusive evidence therefore is not and may never be available since such clouds may not exist. The indication that grains play the decisive role in star formation comes rather from the signs that radiation is an essential factor.

In our Galaxy, apparently as in other spirals, a substantial proportion of the stars are old and are distributed throughout an ellipsoidal volume of moderate flattening; it is now widely accepted that they were formed in an early 'collapse' phase in the evolution of the Galaxy before most of the gas was able to contract to a thin disc. This phase lasted perhaps several 10^8 yr, possibly 10^9 yr; the consequential ultimate distribution of stellar motions has been the subject of many investigations during recent years through the study of 'violent relaxation'.

At the beginning of this phase the volume occupied by the Galaxy was $\gtrsim 30$ kpc across. The temperature of the gas was most probably $\lesssim 10^4$ K and the time taken for a pressure wave to cross it would have been $\gtrsim 2 \times 10^9$ yr. With a magnetic field of $\lesssim 10^{-6}$ gauss and a mean gas density $\lesssim 1$ cm^{-3} the time for a hydromagnetic wave to cross it was $\gtrsim 10^{10}$ yr.

For star formation to have occurred essentially simultaneously throughout the volume, in a time scale of order several 10^8 yr, the extremes of the Galaxy must have been in communication with each other by a means having a transit time of order 10^8 yr or less. Only radiation, with a transit time ~ 10^5 yr, satisfies this requirement.

A similar situation occurs in regard to clusters of galaxies. Many are remarkable for being populated by a preponderance of galaxies of one type; for instance, the Coma cluster consists predominantly of ellipticals. Whatever the differences in the physical conditions necessary to produce different types of galaxy, the uniformity of such conditions across the volume of a cluster of galaxies in a time interval short in comparison with the Hubble time, for any realistic values of gas density and magnetic field strength, again can be obtained only if the initial conditions are determined by radiation, and not by gas pressure or Alfvén waves.

Although interstellar gas interacts directly with the local and general radiation fields, particularly in the case of molecular excitation, the radiation field probably exerts a much more critical influence on the gas indirectly through its effect on grain temperature, and hence on the formation of molecules and solids on grain surfaces and on the thermal interaction between gas and grains. The remarkable discoveries in recent years of interstellar molecules with high abundances have shown them often to be intimately related to regions of recent star formation.

Perhaps it is fair to conclude that there exists a strong suspicion that the interstellar grains and radiation field may be the decisive and controlling factors in the processes of star formation and the evolution of galaxies of stars. Indeed even if there was no observational evidence, it now appears to be such an evident possibility that it is worthwhile examining in some detail, alongside the alternatives. This is done in the second part of the book; the first part is devoted to summarising data which appears to be especially relevant to the problems of star formation.

So far as is possible the subject matter has been divided into separate topics, such as interstellar cloud structure, rate of star formation and so on, and separate chapters given to them for both data and theory. The interrelations between the topics are then considered in further chapters.

This division has been made because it has been found to be convenient in University lecture courses to teach the observational basis in one series of lectures and the theories in another. It is easy to combine the parts of the book if that is required.

Chapter 2.

LOCATIONS OF STAR FORMATION.

The places in which stars are forming at the present time may be located by examining the distributions of very young stars. Given the position of such a star, the precision with which the position locates the place of formation depends on the age and speed of motion of the star. We immediately face a difficulty, in defining the age of the star. Age may be defined as the time elapsed since the protostar became sufficiently dense to be largely unaffected by external anisotropic pressures and so moved freely in the gravitational field. This is the dynamical age. There are other measures of age; for examples; a physical age beginning when the star first reaches a state of equilibrium in which gas pressure balances gravitational attraction near to the main sequence; or when nuclear reactions begin to supply most of the energy requirements; or a rotational age, measured from the time when the star has lost sufficient angular momentum to be able to contract to the equilibrium state.

Highly luminous main sequence stars have short main sequence lifetimes and consequently small physical ages. The dynamical ages may be substantially greater. Suppose, for example, that a fragment of an interstellar cloud contracts to become a protostar, and remains in this state for 10^7 yr before contracting to stellar density; and suppose that it is moving with a random motion of 5 km/s; it will thus have moved 50 pc from the parent cloud before it becomes a star.

Young stars therefore locate the places where protostars become stars, and these may have to be distinguished from the places where the protostars themselves formed.

The first part of this chapter will be concerned with the distributions of young stars; the second part with the distribution of globules and cloud fragments, which may be those from which stars condense; and the third part will examine the relationship between the distributions of stars and interstellar clouds. Evidence indicating the locations of star formation in earlier epochs and in other galaxies will be discussed.

2.01 The Distribution of Young Stars

Main sequence stars of high luminosity including the exciting stars in HII regions are massive and short-lived. Their positions therefore locate the places of formation of massive stars. On the other hand, T Tauri stars are believed to be stars with masses similar to the Sun and in the process of contraction towards the main sequence, and their distributions therefore mark the places of formation of low mass stars.

Compact HII regions appear to be regions active in star formation, and the OH emission sources often associated with them, as well as some of the infrared objects and Herbig Haro objects, may be protostars. Flash stars must also be considered to be possible relatively young stars.

2.011 OB and Wolf-Rayet Stars

Highly luminous main sequence stars, in particular OB and Wolf-Rayet stars, have main sequence lifetimes limited by energy supply to $\lesssim 10^7$ yr, and random motions averaging 5 km/s. They therefore locate their places of origin to within $\lesssim 50$ pc. However, uncertainties in the absolute magnitudes to be assigned to them, and in the corrections to be made for interstellar obscuration, result in uncertainties in determinations of their distances amounting to several hundred parsecs. It follows that the angular distributions of these stars give fairly precise information of the angular distribution of the places of star formation, but the distribution in depth is uncertain. The angular distribution may also lack information in that we do not know how many young stars may be hidden from view by dense interstellar obscuration.

2.0111 Angular distribution in the Galaxy   The gross features of the angular distribution of OB stars as discerned from the northern hemisphere surveys are shown by the surveys of the Hamburg and Warner Swasey Observatories, reproduced by Sim (1968a). The distribution in the southern sky has been given by Geyer (1966), Klare and Szeidl (1966) and Klare (1967). Apart from the sharp concentration towards Cygnus at $l^{II} = 76^{\circ}$, and the broader maximum in Perseus, $100^{\circ} \leqslant l^{II} \leqslant 140^{\circ}$, a noticeable feature of these surveys is the relatively low number of stars in the anticentre quadrant $l^{II} > 140^{\circ}$. On the basis of the OB star distribution alone this could be interpreted as the general level above which the Perseus and Cygnus concentrations rise; but several Galactic features show a marked discontinuity at $l^{II} = 140^{\circ}$ (Reddish, 1967a), Fig. 2.01. This phenomenon has been interpreted as marking the end of the Perseus spiral arm but it may also be caused by a nearby dark cloud (Brück et al, 1968; Dodd, 1974).

The cut-off at $l^{II} = 140^{\circ}$ is complete in the case of the WR stars. If they were distributed similarly to the OB stars, about 10 WR stars would be expected in the range $140^{\circ} < l^{II} < 220^{\circ}$; the probability that their absence is due to chance is only one in twenty thousand (Sim, 1968c).

2.0112 Distributions in depth in the Galaxy   The uncertainties in the distances of young stars, due to uncertainties in their intrinsic luminosities and the amounts of interstellar obscuration, prevent any detailed examination of their distribution in depth. The best that can be done is to determine whether or not they form large-scale structures such as spiral arms. Rohlfs (1967) has shown that OB stars in general, and HII regions and B_e stars in clusters, delineate spiral arms; but that classical cepheid variables and WR stars do not; it is pointed out that in the case of the WR stars this may be due to inadequate knowledge of their distances because uncertain absolute magnitudes and a more recent analysis of new data on WR stars by Smith (1968) gives some indication that they are located along spiral arms.

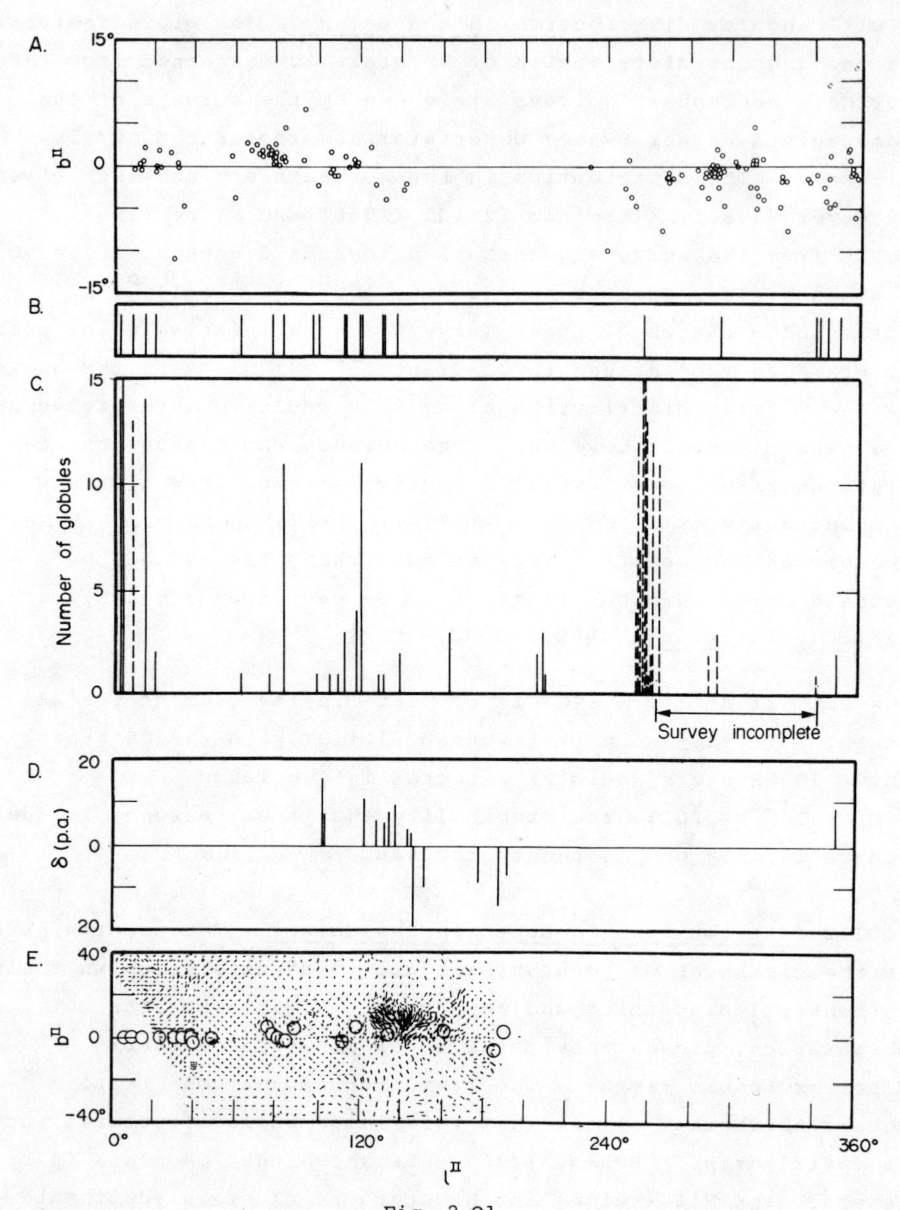

Fig. 2.01
The distributions with galactic longitude of (A) Wolf-Rayet stars; (B) associations and young clusters containing dust-embedded stars; (C) globules (broken lines indicate provisional values); (D) the rotation with wavelength of the position angle of the polarization vector of starlight, from $1/\lambda = 3.04\mu^{-1}$ to $1/\lambda = 1.05\mu^{-1}$; (E) polarization vectors of galactic radio noise at 408 MHz.

The distributions of these various types of star, projected on to the Galactic plane, are shown by Becker (1964), Fig. 2.02.

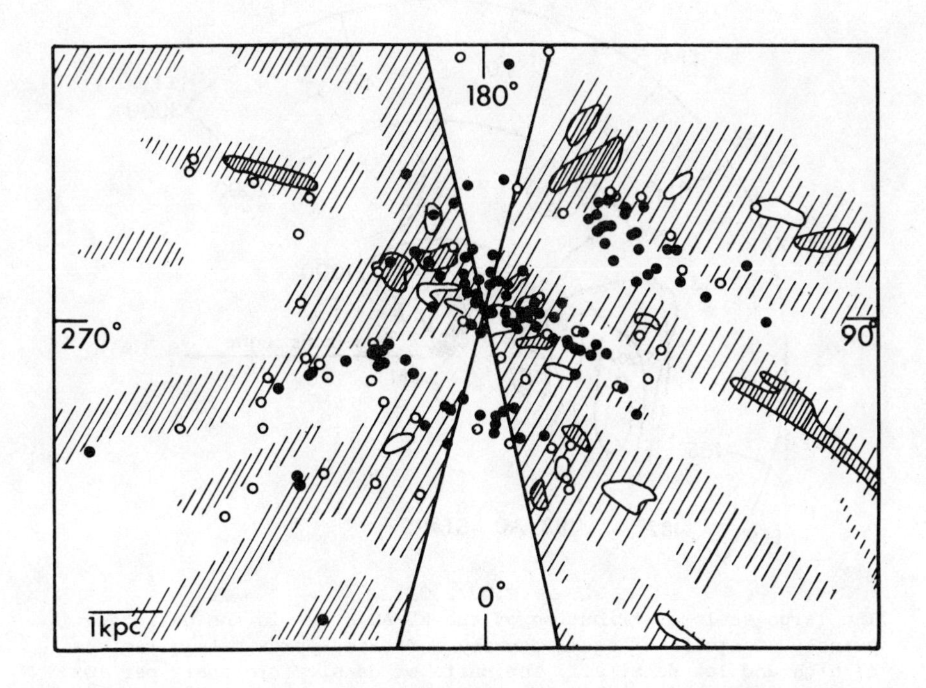

Fig. 2.02
Distributions projected onto the galactic plane of young galactic star clusters and HII regions (filled circles), and of high luminosity δ-Cephei stars (open circles) relative to HI regions (shaded). The sun is at the centre.

The distributions in depth of B5 stars and of B8-AO stars, have been derived from observations in sample areas by McCuskey (1956). Interpolation between the areas gave the contour maps reproduced in Figs. 2.03 and 7.03.

Fig. 2.02 shows clearly the apparent end of the Perseus spiral arm, and Figs. 2.03 and 7.03 indicate that there is a corresponding discontinuity in nearer stellar distributions. As noted above it has been shown that this structural discontinuity is also a feature of the interstellar medium, and of the distributions of globules and dust-embedded stars.

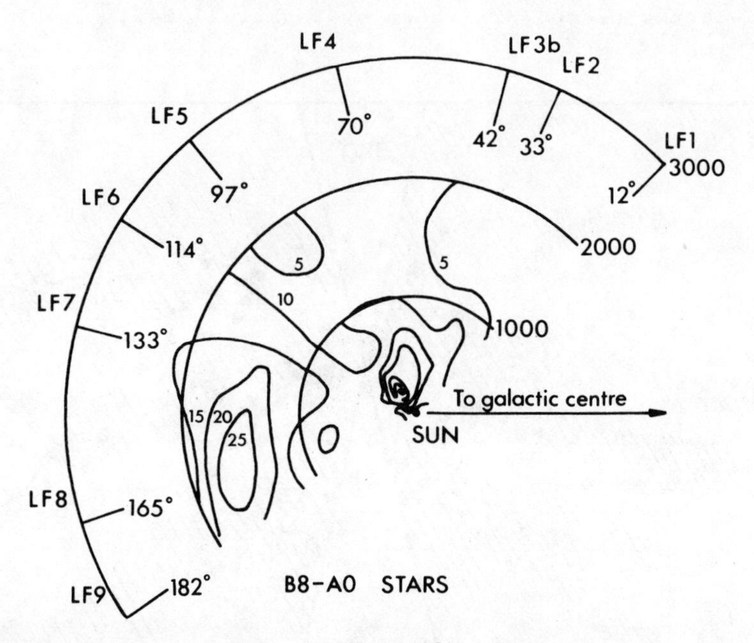

Fig. 2.03
The large-scale distribution of the B8-AO stars in the galactic plane. Contours of equal density indicate approximately regions of high and low density. The units of density are stars per 10^5 cubic parsecs.

Smith (1968) has shown that the distribution of WR stars possesses some remarkable features. Taking the distance of the Sun from the centre of the Galaxy to be 10 kpc, all WC9 stars are within 7 kpc of the centre, most WN6 and WC7 are within 9 kpc, and most other types are more than 9 kpc from the centre. The types concentrated towards the centre, WC6-9 and WN6, are absent from the Magellanic Clouds.

Smith suggests that the reason for this distribution is that the chemical composition of the interstellar material from which the stars form changes with distance from the Galactic centre (evidence of such changes is summarised in section 7.07), and that this changes the subsequent evolution of the stars. The WR stars are supposed to have evolved from stars with masses $\geqslant 25\ M_\odot$ which, having synthesised all the hydrogen to helium in their convective cores through the CN cycle, have lost the outer

15 M⊙, usually by transfer to a close companion (Reddish, 1957; 1974; Paczynski, 1967; Kippenhahn, 1969). We then have a helium star of 10 M⊙ in convective equilibrium, the convection mixing the now high nitrogen abundance into the atmosphere; a WN star. α capture and β decay convert the N^{14} to O^{18} and ignition of the triple α reaction produces C^{12} and a WC star.

We may then imagine that the concentration of WC stars towards the Galactic centre, in contrast to the concentration of WN stars towards the outer parts of the Galaxy and into the Magellanic Clouds, occurs either because they lost sufficient mass to uncover the nitrogen-rich helium core in the first instance, or because mass loss was delayed until after the triple α reaction had been in operation for some time. In either case the implication is that the envelope expansion which leads to the mass transfer or loss is less towards the centre of the Galaxy. Smith suggests that a difference in heavy element abundances is responsible.

It appears, then, that WR stars are the remnants of stars which had a small enough mass to be stable during hydrogen burning in their convective cores, but too large a mass to be stable with essentially pure helium cores; that they are in an advanced stage of evolution but have short lifetimes. The main sequence lifetime of a star of mass 25 M⊙ is about 5×10^6 yr. If this interpretation of the evolutionary status of WR stars is generally correct then they are essentially young objects comparable with the O stars, although Mikulasek (1969) argues that they are probably older than the O-associations.

2.0113 Distributions in other galaxies

(a) Magellanic Clouds The distribution of young stars in both of the Clouds has been determined by means of two-colour composite photography (Walker, Blanco and Kunkel, 1969).

In the Large Magellanic Cloud (LMC), the red giants are concentrated into the bar, whereas the OB stars and red supergiants are widely and fairly uniformly distributed. Making the reasonable assumption that the red giants are relatively old stars, it is concluded that star formation occurred first in the bar but is now taking place more or less uniformly throughout the LMC.

In the SMC on the other hand, young and old stars are distributed similarly and star formation appears to have taken place fairly uniformly in space at all times. The uniformity in the distribution of young stars in the SMC at the present time is shown also by the objective prism survey by Sanduleak (1968, 1969).

(b) <u>M33</u> Two-colour composite photographs of M33 (Walker 1964) show the remarkable concentration of OB stars into spiral arms embedded in a smooth spheroidal distribution of older red giants. The spiral arms reach into the nucleus. In these central regions the ratio of red supergiants to OB stars is less than in the outer parts, which may indicate that star formation has occurred more recently towards the centre, or that the physical properties of the stars formed there are different.

(c) <u>M31</u> The bright stars in M31 are concentrated into four arms (Reddish 1962c) intersecting the major axes at angular distances from the centre of 41, 47, 56 and 95 arc minutes. The widths of the bands of stars are about 3, 4, 6 and 18 arc min respectively to half peak density; the densities peak very sharply in the first two of the arms, showing a high concentration of stars in a width of 1 arc min. The third arm is less sharply peaked and the fourth has a broad maximum. At the distance of M31 (assuming a true distance modulus $24^{m}.2$), 1 arc minute corresponds to 200 pc.

2.0114 <u>Aggregation</u> Analysis of the distribution of OB stars led Roberts (1957) to conclude that all OB stars form in associations and clusters. It is, however, difficult to draw reliable conclusions from the apparent distribution of OB stars on account of the patchiness produced by dense obscuring clouds, and the large errors in estimates of their distances. The

Hamburg-Warner-Swasey survey of luminous stars in the Milky Way reaches to $m_{pg} \approx 13$ and provides more extensive data on the distribution of OB stars than was available to Roberts.

The stars in the survey are divided into three classes OB^+, OB, OB^- in order of decreasing luminosity. The OB^+ are almost entirely O stars, the OB average O9, the OB^- average B1 to B2, possibly reaching as late as B5. A superficial examination of the distributions plotted by Sim (1968a) does not indicate any greater tendency to clustering among the OB^+ than among the OB or the OB^-.

The average ages (half main sequence lifetimes) are $\approx 3 \times 10^6$ yr (Hoyle, 1960). With an r.m.s. velocity dispersion of 9 km s^{-1} (Allen, AQ2, p.243) the average distances moved by the OB stars from their places of origin is 27 pc. With an average line-of-sight distance of 1.5 kpc (limited by obscuration) the corresponding average angular motions are $\approx 1^o$. It is possible that such motions would be sufficient to produce the observed spreading of the majority of OB stars if they were indeed formed in associations. Some at least of the observed patchiness is due to obscuration; a minority of OB stars are clearly clustered; the degree of patchiness may be evidence that OB stars form in aggregates whose dimensions are comparable with the thickness of the Galactic plane distribution for an extreme population I system, as if the spiral arm is a string of blobs; but even then there is a considerable substratum of stars which appear to have formed 'in the field' rather than in the large aggregates.

To summarise, a minority of the OB stars are clearly clustered, a substantial number are widely and apparently randomly scattered, and another substantial number show a large-scale patchiness some of which is due to holes in the obscuration but the rest of which may indicate real OB star aggregates. Possibly a majority of OB stars form in such aggregates, but there is no compelling evidence that all of them do, or that typical large aggregates represent anything more than a random distribution of stars in a region in which there has been for a considerable time more than the average amount of material out of which stars form. In other words, there does not appear to be strong evidence that most stars form in aggregates where an aggregate represents the condensation of a cloud into stars at a particular epoch. Indeed, from an examination of the motions of eight stars which may have formed in and escaped from clusters or associations, Aveni and Hunter (1967) conclude that four of them most probably formed as isolated stars, and Steffey (1968) considers that several high-latitude supergiants probably formed in isolation, some of them away from the Galactic plane.

The concept of expanding OB associations (Ambartsumyan, 1949; Blaauw, 1964) led to the possibility that OB associations are the seat of origin of all OB stars, but the absence of a dependence of the number density of stars in associations and clusters on the ages of the systems (Reddish, 1967c) is in conflict with the evidence of expansion and dispersion. Kholopov (1969) has concluded that many comparatively compact O-associations are not dispersing but are gravitationally bound.

The widespread distribution of WR stars in the LMC (Westerlund and Smith, 1964) also indicates widespread formation of stars.

It is observed in our Galaxy and in the LMC that field WR stars are fainter than those of similar type in clusters and associations (Smith, 1968; Reddish, 1968b). Possibly this may be taken to indicate that the field WR stars are older, and are ones which were formed in aggregates and have moved into the general field; although they could not have travelled the distances they are observed to be from aggregates, within the lifetimes set by the evolutionary considerations discussed earlier, unless they have higher velocities than OB stars. There is, however, another interpretation which can be put on these observations other than an age effect. It is that the masses and luminosities of the brightest stars formed increase with the total masses of the aggregates in which they form. This would not be surprising because, in a fragmentation process having a mass spectrum for the fragments similar to the stellar initial mass function, there is evidently a statistical upper limit to the mass of the largest fragment formed (if the lower limit is determined by physical processes) which increases with total cloud mass (see section 3.041).

2.0115 Aggregation in other galaxies

(a) Magellanic Clouds Several of the globular clusters in the LMC contain large numbers of OB stars and consequently are apparently young clusters (Woolley, 1960). Walker *et al* (1969) have noted that these blue clusters and the OB associations are strung out along several narrow parallel curved lines extending across the bar from the region of the 30 Doradus nebula.

The fewer OB clusters and associations in the SMC are fairly uniformly scattered over the area of that galaxy, it being notable that the outer 'ring' of the galaxy appears to contain only such a young population.

(b) M33 The grouping of the OB stars into aggregates is very noticeable in the composite pictures produced by Walker (1964), who draws attention to the similarity between many of the groups and the Galactic OB clusters h and χ Persei. Nonetheless, besides the large number of aggregates there is evidently also a widespread distribution of very small groups or single OB stars. The aggregates appear to occur more closely to the centre lines of the spiral arms than do the single stars.

(c) M31 A catalogue of 188 OB associations in M31, detected by multicolour photographic photometry, has been given by van den Bergh (1964). They are generally large in comparison with Galactic associations; this may be in part a selection effect, but van den Bergh points out that a number of the M31 associations have dense cores surrounded by older envelopes, and that the cores have dimensions comparable with Galactic associations. Kholopov (1969) has concluded that galactic clusters and associations have a core-halo structure which largely represents the distribution of the stars at the time they were formed.

The main concentration of OB associations in the SW sector of M31, van den Bergh's Figs. 2 and 4, coincides with the high peak in the distribution of bright stars in that region (Reddish, 1962c) and generally the associations occur along the arms delineated by the counts of bright stars.

It should be possible from observations such as these to estimate the proportions of OB stars which occur in aggregates, but it is difficult to compare different surveys to very faint magnitude limits on account of uncertainties in what those limits are. However, van den Bergh estimates that the total number of OB associations in M31 is 200, and the richest of them, OB78 (van den Bergh, 1966), contains 331 stars brighter than B = 21. Therefore it appears that the total number of OB stars brighter than B = 21 in associations in M31 is $< 7 \times 10^4$. On the other hand, the total number of all stars in M31 brighter than $M_{pg} = 21.1$ has been found to be 1.2×10^5 (Reddish, 1962c), from which it would appear that not more than half the bright stars in the galaxy occur in associations and the fraction may be considerably less than this.

We are therefore still faced with the possibility that stars can form singly in the general field, as well as in aggregates.

2.0116 Distribution within aggregates The existence of subgroups within OB associations has been discussed by Blaauw (1964), who notes that, within associations, the proportion of earliest type stars is always largest in the most concentrated subsystem. Also the main sequences indicate increasing age with increasing dispersion of the subsystem. These observations support the view that aggregates of OB stars are dispersing into the general field; evidence to the contrary has been quoted in 2.0114.

The concentration of OB stars into clusters within associations is marked in several cases, for example, in Cep OB4 (MacConnell, 1968), in Ori OB1 (Blaauw, 1964) and in Per OB1 - the double cluster. On the other hand some associations, notably Cas OB5, have an annular distribution of stars (see the OB star distribution given by Sim (1968c), and it has been suggested that many OB stars form in shells, thereby presenting the appearance of rings of stars (Isserstedt and Schmidt-Kaler, 1968). Ambartsumyan has drawn attention to chains of OB stars in our own and other galaxies.

No clear single pattern emerges to describe the spatial distribution of OB stars formed in aggregates.

2.012 <u>T Tauri Stars</u>

2.0121 <u>General Distribution</u> An extensive summary of the distribution and properties of T Tauri stars has been given by Herbig (1962). The main features of the distribution are summarised as follows.

T Tauri stars are generally found only in regions where there is dense and extensive obscuration, and they are always in, or close to, bright or dark nebulae. Often they are concentrated in the vicinity of bright stars embedded in dark nebulae.

On the other hand, regions of dense obscuration do not necessarily contain many T Tauri stars, and they are rarely found in small, sharply defined dark clouds. None have been found in Bok's globules.

On account of their intrinsic faintness, and their predominant occurrence in obscured regions, the distribution of T Tauri stars is known only within several hundred parsecs of the Sun. We have no knowledge of the general galactic distribution, excepting the dispersion perpendicular to the Galactic plane which in the solar region of the Galaxy is $|\bar{z}|$ = 53 pc (Kholopov, 1960b).

2.0122 <u>Aggregation</u> The major known concentrations of T Tauri stars are in the region of Orion (1000), NGC 2264 (500) and NGC 7000 (500), the total numbers having been estimated from sample surveys (Herbig, 1962).

The groupings of T Tauri stars in the so-called T-associations have been considered particularly by Kholopov (1958, 1960a, 1960b). T-associations often occur in groups, distributed in the form of straight or curved chains. The associations are sometimes elongated in the direction of the chain. The peak surface densities of T Tauri stars in these systems, which have dimensions of some tens of parsecs, occur where the obscuration is highest, the concentrations therefore being real, not apparent.

All O-associations near enough to be searched for T-associations are found to contain them; 8 out of 29 definite T-associations are in O-associations. The T-associations are always involved with nebulosity. On the other hand, not all T-associations have O-associations connected with them, but those that have are the richest in the number of T Tauri stars. The distribution of spectral types among the T Tauri stars is the same in those T-associations that are connected with O-associations as in those that are not.

T-associations which do not have nebulae intimately involved with them are generally larger and less dense, and Kholopov suggests that this may indicate that these are older systems which have expanded from a more dense distribution.

2.013 HII Regions

2.0131 Distribution in the Galaxy Catalogues of Galactic HII regions have been given by Murdin and Sharpless (1968) and by Courtes *et al*, (1968). The former give data for 61 HII regions, including distances derived from spectroscopic parallaxes of exciting stars, diameters, and densities from the dependence of Stromgren radius on density, assuming that the HII regions are ionisation limited. This assumption is supported by the analysis of Higgs and Ramana (1968). Courtes *et al* give radial velocities determined by interferometry for 187 Galactic HII regions, and distances determined by use of a model for Galactic rotation for 101 of them.

Only a minority of HII regions detected by their emission at radio wavelengths have been identified with visible objects. Altenhoff (1968) has noted that of 176 HII regions detected by 2.7 GHz continuum radiation, only 19 definite and 21 probable identifications can be made with visible objects. This is presumably an indication of heavy obscuration by interstellar dust in regions of star formation.

Murdin and Sharpless (loc. cit.) show that the average distance of visible HII regions from the Galactic plane is $|Z| = 50$ pc, similar to that of OB stars and T Tauri stars.

The distribution of HII regions with distance from the centre of the Galaxy has been described by Burke (1968), on the basis of information derived from radio wavelength continuum observations. There are some HII regions within the Galactic nucleus, but there are few within the expanding region which extends from outside the nucleus to 4 kpc from the centre. At that distance, where expansion stops and circular rotation predominates, a major concentration of giant HII regions begins and extends to 6 kpc. Beyond 6 kpc from the centre there are relatively few giant HII regions, a secondary condensation occurring at about 8 kpc and the strongest individual HII region of all, W49, appearing in isolation in the outer regions of the Galaxy. HII regions in the vicinity of the Sun are smaller and presumably typical of a widespread distribution. Again the observations indicate the great extent of the obscuration by interstellar dust of regions of star formation.

It is presumed in this argument that the giant HII regions are excited by OB stars, or at least by processes of star formation.

The implication of the distribution is that the region between 4 kpc and 6 kpc from the centre of the Galaxy is most active in star formation. We need to consider why this should be so, and whether or not it is a permanent feature of the distribution of star formation within the Galaxy. Observations of the distributions of HII regions in other galaxies are pertinent.

2.0132 Distribution in other Galaxies A description of the spiral arm structure and the distribution of HII regions in the galaxy M31 has been given by Baade (1963), Baade and Arp (1964) and Arp (1964).

Spiral arms appear at distances of 3 and 10 arc min from the centre, and then every 20 arc min approximately to 110 arc min. Numbering the arms from the centre outwards, arm 1 contains dust but no HII regions or bright stars; arm 2, dust and a few bright stars; arm 3, dust, bright stars and HII regions. The greatest numbers of OB stars and HII regions occur in arms 4 and 5, which are less dusty; arms 6 and 7 have a few OB stars and no dust. The general feature of a maximum halfway from the centre to the edge is similar to that in our own Galaxy. To that extent there is agreement with the overall distribution of bright stars (Reddish, 1962c), but the detailed correlation between Baade's arms and those determined from the bright star counts is not good. The cause of this may lie at least in part on the different amounts of obscuration suffered by OB stars observed in the blue region of the spectrum, and Hα observations of HII regions, although bright star counts were also made on the Hα sensitive plates and agreed well with the blue sensitive counts.

Hodge (1969a) has examined the radial distribution of HII regions in 25 spiral galaxies, and finds that on average the number density peaks at one quarter of the distance to the outermost HII region. Beyond the peak the number declines exponentially with distance.

The distribution in irregular galaxies (Hodge, 1969b) is found to be asymmetric, and Hodge suggests that the distribution may indicate bursts of star formation occurring in regions with a characteristic size of 1 kpc.

Similar conclusions may be drawn from the observations of a variety of galaxies by Morgan (1968). In spiral galaxies the

HII regions are concentrated into the spiral arms but the asymmetries are sometimes remarkable - for example, one spiral arm of a galaxy having many HII regions and the diametrically opposite arm having none.

These observations indicate that star formation occurs regionally, in bursts at intervals long in comparison with the lifetimes of HII regions, that is at intervals >> 10^7 yr. It is not evident whether star formation regions migrate by triggering star formation in neighbouring regions, or develop spontaneously at unconnected locations.

2.0133 Stellar aggregates in HII regions It was noted in 2.0122 above that, in the region of the Galaxy between 4 kpc and 6 kpc from the centre, interstellar obscuration is so high that the giant HII regions detected by their radio emissions cannot be seen at optical wavelengths. Baade and Arp (1964) point out that M31 similarly contains a great number of very heavily obscured emission nebulae.

In compact HII regions (Mezger *et al*, 1967) the obscuration is so high that the most intense parts of the regions, as indicated by the contours of continuum emission at radio wavelengths, often are not visible in Hα light, and except in Orion the exciting stars have not yet been observed (Schraml and Mezger, 1969). It must be concluded that some ten or more magnitudes of obscuration are located in them.

Mezger *et al* (loc. cit.) and Mezger and Robinson (1968) have provided convincing evidence that compact HII regions are associated with groups of protostars and newly formed stars. About 8 O-stars are required to account for the emission from the region W49, (Mezger *et al*, 1967), and more than 50 O-stars are probably present in W51 (Terzian and Balik, 1969). The compact HII regions, and the heavily obscured zone of giant HII regions between 4 kpc and 6 kpc from the Galactic centre, therefore probably indicate the places in which aggregates of O-stars are forming. The bright HII regions observed in other galaxies, as described in section 2.0122, presumably also relate to aggregates of O-stars, not to individual stars.

2.014 OH Anomalous Emission Sources

Intense OH emission comes from regions of small angular diameter in about a third of all HII regions (Mezger and Robinson, 1968; Turner, 1970), predominantly the compact HII regions. There is close coincidence in the radio velocities of the OH and HII emission sources as well as in their apparent positions, and there can be little doubt that we are observing small dense clouds of OH molecules, embedded in a large cloud of hydrogen excited by newly formed O-stars. Clusters of several OH condensations are usual, interferometer measurements giving diameters of $\sim 10^{14}$ cm or less for the individual condensations. Menon (1967), Mezger et al (1967), Shklovsky (1967), and Mezger and Robinson (1968), have argued that these objects are protostars, probably of smaller mass than O-stars.

If the anomalous OH emission sources are indeed protostars, the fact that they generally occur in groups embedded in compact HII regions cannot be taken to indicate that such protostars only occur in groups, because it may be that the presence of radiation from newly formed nearby stars is necessary to excite the OH emission (Litvak, 1969). Furthermore Ball and Staelin (1968) have shown the characteristics of relative line strength and polarization in the OH source emission to be related to the properties of the HII regions in which they occur; in particular to whether or not non-thermal sources, probably supernovae remnants, occur in those regions.

That the OH sources may be protostars of relatively small mass is supported by the observation that OH sources are absent from a number of systems that show few or no T Tauri stars, for example, NGC 2024 and NGC 2244 (Menon, 1967), and by calculations of the total masses of the groups of OH sources from their radial velocity distributions on the assumption that the groups are gravitationally bound (Shklovsky, 1968). OH emission has been detected from the regions of two T Tauri stars in Taurus and Auriga which are observed to be in isolated dust clouds (Gahm and Winnberg, 1971). Since the sources of anomalous OH emission generally occur in the compact HII regions it would follow, if these arguments are correct, that the formation of stars of large mass is usually accompanied by the formation of low mass stars; but if the radiation from stars or protostars of large mass is needed to excite the OH emission, it does not follow that low mass stars are not also formed elsewhere, in regions where stars of large mass do not or have not formed.

In a systematic search of an area of the sky for OH emission sources not necessarily connected with HII regions, (Ellder et al, 1969), at least one source has been found in a region which does not show thermal radio emission, and the four sources found all appeared to be in places where the red colours of the stars indicate high interstellar obscuration. Crutcher (1973) from a survey of 62 clouds concludes that all dark dust clouds have OH molecules, and Verschuur (1971) has detected normal OH emission from several HI clouds not having an obvious dust content.

2.015 Infrared Sources

It is widely agreed in the literature that some of the infrared sources are protostars or are associated with newly formed stars.

The distribution of infrared sources in the Galaxy has been examined in surveys by Neugebauer and Leighton (1968) and by Ackermann (1970). The surveys show that whereas the brighter infrared stars are well distributed over the sky, the fainter ones are strongly concentrated towards the Galactic plane. Ackermann shows that they are also strongly concentrated towards dark nebulae and in some cases towards bright nebulae. It is not yet evident whether this represents the reddening effect of interstellar obscuration, or the existence of larger numbers of protostars in dark clouds and in regions containing newly formed OB stars.

2.016 Dust-Embedded Stars

An examination of the distributions of interstellar obscuration of stars in 66 OB clusters and associations, based on published UBV photometry, showed that in 22 of them the obscuration varies considerably from star to star and increases significantly with stellar luminosity (Reddish, 1967c). It was concluded that much of this obscuration must occur in circumstellar clouds.

Using the main sequence lifetime of the brightest main sequence star as a measure of the age of each system, it was found that circumstellar dust clouds always occur in systems with ages $\leqslant 10^5$ yr, sometimes in systems with ages between 10^5 yr and 2×10^6 yr but never in systems older than 2×10^6 yr. Therefore it appeared reasonable to conclude that the circumstellar dust clouds are the remnants of the clouds out of which the stars condensed and are indicative of very young star systems. This investigation followed earlier ones by Blanco and Williams (1959) of Cepheus IV, and by Hoffleit (1953, 1956) who obtained indications of circumstellar clouds around field OB stars. If this were confirmed it would be strong evidence that these stars formed in isolation.

The distribution of the 22 clusters and associations containing dust-embedded stars is similar to that of the Wolf-Rayet stars - they are concentrated into the Galactic plane but are totally absent from the region with Galactic longitudes $140^\circ \leqslant l^{II} \leqslant 230^\circ$ Outside this region, Wolf-Rayet stars are often found in systems which contain dust-embedded stars, rarely in those that do not.

Dust-embedded stars show no systematic tendency to group together more closely than other stars in the systems within which they occur although such groupings do sometimes occur. The association Cygnus OB2, for example, has a large number of dust-embedded stars; the four most luminous are also the four most heavily obscured and they include the three Wolf-Rayet stars in the association (Reddish *et al*, 1966; Reddish, 1968b).

These four stars are widely spread over the area of the association. On the other hand, the dust-embedded stars in Ceph OB4 are grouped preferentially together within the association (MacConnell, 1968).

Subsequently, a variety of investigations have confirmed the existence of such circumstellar dust clouds around OB stars (Schmidt, 1971; Strom et al, 1972; Allen, 1972; Riegel and Crutcher, 1972a; Kemp, 1974), B_e and A_e stars (Strom et al, 1972), A and F stars (Strom et al, 1971), and infrared objects (Cohen, 1971; Kellerman and Pauliny-Toth, 1971; Penston and Allen, 1971, and Wilson, 1971). In addition several theoretical studies have shown that when a star forms, a substantial part of the mass of the protostellar cloud is left behind as a circumstellar shell during the contraction (Ireland, 1968; Dorschner, 1971; Baglin et al, 1973; Appenzeller and Tscharnister, 1974; Berruyer, 1974; Kolesnick and Nadyozhin, 1974).

2.02 The Distribution of Interstellar Gas, Clouds and Globules

It is not certain that the small dense roundish dark clouds, which were denoted globules by B.J. Bok, will condense to form stars rather than evaporate or disperse. If they do indeed form stars, they may represent a phase between the fragmentation of a cloud and the formation of a protostar which is of especial significance in that they may represent the earliest stage at which the process of formation of a single individual star can be observed. They will therefore be in the nature of an interface between interstellar clouds and newly formed stars.

Our knowledge of the distribution of interstellar gas in general and of clouds in particular has been improved enormously by the data resulting from 21 cm hydrogen line surveys of our Galaxy, and by observations of HII regions at both radio and optical wavelengths. These, and similar observations of other galaxies, have also shown relationships between the distributions of atomic hydrogen and young stars. However, the growing indirect evidence that much of the interstellar hydrogen may be in molecular form, especially in dark clouds, and that stars probably form from such clouds amidst very high obscuration by grains, means that we are almost wholly ignorant of the true relative galactic distributions of young stars and of the kinds of clouds from which they form.

The following sections will deal with the general distributions of HI and HII, cloud structures, clouds of molecules, dust clouds, and globules.

2.021 The General Distribution of Interstellar HI and HII

2.0211 General distribution in the Galaxy Extensive surveys of the 21 cm hydrogen line emission (Muller and Westerhout, 1957; Westerhout, 1957; Schmidt, 1957) indicated that the interstellar neutral atomic hydrogen in the Galaxy is concentrated into a pair of spiral arms having about six turns and extending to about twice the distance of the Sun from the centre of the Galaxy. An extensive summary of the data has been given by Kerr (1969). Beyond 4 kpc from the Galactic centre (assuming $r_{\odot}$ = 10 kpc) the predominant motion of the gas is circular, but within 4 kpc it is radial, and within 1 kpc from the centre it becomes circular again.

A new interpretation of the HI distribution has been made recently by Courtes *et al* (1969) based on the close correspondence between the radial velocities of the HI gas and HII regions in the same direction, and using distances to the HII regions derived from spectroscopic parallaxes of the exciting stars. It is found that the spiral arms in the solar vicinity are separated by about 2 kpc and have a pitch angle of 20°, from which it follows that there must be more than just one pair of spiral arms; it is concluded that the Galaxy is a multi-arm spiral of type Sc.

The density distribution of HI in a direction perpendicular to the Galactic plane is gaussian, with a thickness to half-density points which remains remarkably constant at different distances from the centre of the Galaxy as far as the Sun, 80 pc at 0.5 kpc, 110 pc at 4 kpc and 170 at 10 kpc. Beyond the Sun the thickness increases rapidly (McGee and Milton, 1964; Lozinskaya and Kardashev, 1964).

The radial distribution of HI density shows a broad maximum at 12 kpc from the centre and a secondary maximum at 8 kpc. Averaged over distances of about 1 kpc in the plane and the thicknesses perpendicular to the plane given above, peak densities are 0.3 to 0.7 hydrogen atoms cm^{-3}.

The density distribution of HI apparently contrasts with that of HII (Oort, 1965a) which shows a much sharper maximum at 5 kpc from the centre, just outside the expanding region. If the HII emission is excited by OB stars this would indicate that the rate of star formation does not depend predominantly on the average density of HI. It is possible, however, that a great deal of the gas in the inner regions of the Galaxy is in molecular form. Both the HI and HII distributions have secondary maxima coinciding at about 8 kpc from the centre.

Subsequent surveys with increasingly high spatial and frequency resolution have given the distribution of HI in more detail, and these will be described in section 2.022 dealing with cloud structure.

Measurements of interstellar Lyman-α absorption (Jenkins, 1969); of interarm obscuration by dust at visible wavelengths (Reddish, 1968d), and of absorption of soft X-rays (Rappaport _et al_, 1969) show that the interarm density is in the range $0.1 \leqslant n_H \leqslant 0.3\ cm^{-3}$. Since the total amount of HI in the solar region of the Galaxy corresponds to an average space density $n_H \approx 0.7\ cm^{-3}$ it is probable that this mainly represents HI concentrated into spiral arms with average densities $n_H \approx 2\ cm^{-3}$ or higher. This interpretation is supported by the close correspondence between the radial velocities of the peak HI densities and those of HII regions, which delineate spiral arms (Courtes *et al*, 1969) and by the sharp cut-off in the hydrogen line-of-sight densities, with increasing galactic latitude, as noted by Heiles (1967a).

2.0212 *General distribution in other galaxies* The gross features of the distribution of HI in nearby galaxies have been determined by 21 cm line radiation measurements with angular resolutions of several minutes of arc. The HI appears to extend to much greater distances from the centre of a galaxy than do the optical features, although photographs to very low brightness levels (van den Bergh, 1969b) show faint extensions of the outer parts of some galaxies which may cause us to revise earlier conclusions. There is uncertainty also in the degree of correlation between the distributions of HI and HII.

Detailed surveys of the galaxy M31 (Gottesman *et al*, 1966; Davies and Gottesman, 1970; Fig. 2.04) have shown the distribution of HI to be similar to that in our Galaxy, having a deficiency in the central regions and reaching a maximum at about 9 kpc from the centre. Contrary to the apparent situation in our Galaxy, however, the distribution of HII regions appears to follow closely that of the HI.

Similar ring-like distributions of HI in spiral galaxies are reported by Roberts (1968), who notes that in M31, an Sb galaxy, the spiral arms of HII and early type stars are embedded in the ring, but in Sc type systems the ring of maximum HI density lies outside the predominant spiral arm features, as it does for our Galaxy; this accords with Courtes' classification of our Galaxy as Sc, referred to in the preceding section. However, in the case of M33 at least, the central region containing the spiral arms has a surface density of HI which is 90% of the peak in the ring and there do not appear to be central depressions in NGC 2403 (Guelin and Weliachew, 1969) and in many other spiral galaxies (Bottinelli *et al*, 1968); on the other hand in M31 the central depression is about a factor 2 in the surface density.

Irregular galaxies, the Magellanic Clouds in particular, have no central depression in the HI distribution, and the HII and HI distributions are similar (Hindman, 1966; McGee and Milton, 1966).

Lewis (1972) argues that since 21 cm H-line velocity rotation curves of galaxies indicate much mass outside of the luminous body, and also that ρ (HI)/ρ(stars) is increasing outwards, the 'missing' mass may be optically thick HI with a low spin temperature. Such a model for NGC 300 with 87% of the mass in a disc of cold gas gives results in agreement with observation.

2.022 The Distribution of Interstellar Clouds

The physical characteristics of clouds will be considered in a later section. In this section we are concerned only with the spatial distribution of clouds.

The high resolution 21 cm hydrogen line survey by Heiles (1967a) shows the HI cloud structure in the Cygnus-Orion arm of the Galaxy, at a distance of about 500 pc in the region $100^o \leqslant l^{II} \leqslant 140^o$, $13^o \leqslant b^{II} \leqslant 17^o$. About a half to three quarters of the gas is distributed fairly uniformly along the arm, with a density $n_H \sim 1\ cm^{-3}$. The remainder is concentrated into two streams, each about 20 pc across and having an average density about 2 cm^{-3}, embedded in the smooth distribution and extending along the arms over distances greater than those observed. About a fifth of the mass in these streams is in cloudlets with densities $n_H \sim 4\ cm^{-3}$, sizes of a few parsecs, and approximately stellar masses. Several aggregates of such clouds occur.

From a 21 cm H-line absorption study of 97 clouds, Hughes et al, (1971) conclude that in the region of the Sun 75% of the mass is in the cool dense state, the average over the Galaxy as a whole being 40%.

The thickness of the layer of neutral hydrogen in the solar vicinity is 170 pc to half-density height (see section 2.0211) and Heiles (1967a) notes a sharp cut-off at $l^{II} = 28^{o}$ in gas which is estimated to lie at distances 300 pc to 500 pc, indicating a thickness of 150 pc to 250 pc.

On the other hand the dark clouds which must be much more dense if they contain grains in a similar abundance to regions which produce no significant obscuration, are more concentrated to the plane of the Galaxy. The survey of interstellar reddening by Fitzgerald (1969) measures dusty clouds which are transparent enough for stars to be observed through them, and finds thicknesses from 80 pc to 200 pc. The dark clouds mapped by Lynds (1962) are so opaque that no OB stars have been observed through them; the mean densities are $n_H \sim 300\ cm^{-3}$ and average thicknesses $\sim$ 100 pc (Samson and Reddish, 1975). The interstellar hydrogen in these dark clouds is probably H_2 (Heiles, 1967a) since the 21 cm line of HI is not observed either in emission or absorption, and OH and H_2CO molecules are often present in significant amounts (Cudaback and Heiles, 1969; Palmer *et al*, 1969).

The location of the dark clouds relative to the main spiral arms of HI is uncertain because of the difficulty of determining distances in both cases.

The relatively high abundances of interstellar molecules in the direction of the Galactic centre (Whiteoak and Gardner, 1969), where they appear in absorption similarly to, but more strongly than those in dark, cold clouds towards the Galactic anticentre, may indicate that the major concentrations of such dark clouds occur in the inner spiral arms 4 kpc to 6 kpc from the Galactic centre where the maximum concentration of HII regions occurs. Other evidence of very high obscuration in this region has been noted earlier (sections 2.013 and 2.014).

2.023 Globules

Bok and Reilly (1947) drew attention to 16 dark globules of interstellar matter, with angular diameters ranging from 6 arc sec to 1 arc min, seen in projection against the diffuse nebula

M8. About twenty others were noted in Carina, Ophiuchus and Sagittarius, and where the background was bright enough for them to be searched for over large areas of the sky they were as frequent as one per square degree. In that paper and in a subsequent one (Bok, 1948) it was suggested that the globules are protostars.

The requirement of a bright background against which to be able to see globules means that it is not possible to make a complete survey of their distribution in the Galaxy, or to make a survey free from selection effects consequent upon any real spatial relationship there may be between globules and bright nebulae or dense star fields. Furthermore, the difficulty and tedium of making surveys for globules is such that surveys have been restricted to small areas of the sky.

Counts of numbers of globules detected against bright backgrounds have been given by Hoffleit (1953) for 75 square degrees in Carina, and by Roshkovsky (1954) for 21 bright nebulae.

A detailed survey of the numbers and distribution of globules in 57 OB clusters and associations has been made by Sim (1968b). In addition to the position of each globule, the catalogue gives the shape, size and orientation. The globules are found to be preferentially concentrated into those OB star systems which contain the larger proportions of dust-embedded stars, and they are apparently substantially deficient, although not entirely absent, in the galactic anticentre quadrant which is devoid of Wolf-Rayet stars and dust-embedded stars. It is possible that this deficiency is only an apparent one due to the difficulty of detecting globules in a region lacking a bright background. On the other hand, it is also shown by the distribution of small dark nebulae detected by Lynds (1962) from star counts.

2.03 Spatial Relationships between Young Stars and Interstellar Matter

The largest numbers of young stars appear to occur where there are the greatest amounts of interstellar matter from which stars may form. In these regions it also appears that the interstellar matter is most dense.

The evidence for the first of these conclusions comes primarily from observations of other galaxies, on account of the difficulty of determining with sufficient accuracy the distances to stars and gas in our own Galaxy.

2.031 <u>General Distributions</u>

In other galaxies the apparent surface density distributions of young stars and HI are generally well correlated with each other. This is true especially in the LMC (McGee and Milton, 1966), in the SMC (Sanduleak, 1969) and in M31 (Gottesman <u>et al</u>, 1966; Davies and Gottesman, 1970). There may be exceptions to this overall correlation in some Sc galaxies, for which Roberts reports hydrogen extending far out beyond the stars; but measurements by other astronomers and observations of other Sc galaxies produce conflicting results (see section 2.0212).

Some uncertainty in the interpretation of apparent correlations between the spatial distributions of new stars and gas arises through the absence of any direct measure of the amount and distribution of interstellar H_2. In the Galaxy, interstellar OH molecules appear to occur in greatest abundance in dark (dense, dusty) clouds (Turner, 1969a; Cudaback and Heiles, 1969) and in the regions nearer to the Galactic centre than is the Sun where, on the basis of the HII distribution, star formation is judged to be most active (McGee <u>et al</u>, 1967; Oort, 1965a). In view of the evidence that stars form from dark clouds, this latter distribution may indicate a relatively high abundance of molecules, possibly including H_2, in dark clouds in regions where the HII is abundant and the HI density apparently reduced. There are indications that obscuration by dust in the Norma 'arm', at 4 kpc to 7 kpc from the Sun towards the interior of the Galaxy, amounts to several magnitudes at least (Lynga, 1968; Westerlund, 1969). It is therefore not certain that the surface density of interstellar gas in some Sc galaxies has the shallow central minimum shown by the HI alone.

Amounts of dust up to 10^5 M⊙ in the central regions of galaxies which spectra show to be populated by population I objects have been estimated from multicolour photometry (Pronik, 1974).

2.032 Detailed Distributions

If Bok's globules are protostars (Bok, 1948), a possibility which is still a subject of debate (Herbig, 1970; Reddish, 1970), they probably represent the earliest stage in the formation of stars yet observed. The anomalous OH emission sources, which are also considered to be protostars (Shklovsky, 1966, 1969; Mezger and Robinson, 1968), may then represent a subsequent stage in which an infrared protostar has formed in the core of a globule. Some of the infrared sources may be a later stage of development (Davidson and Harwit, 1967) although many of them are possibly evolved red supergiant stars with circumstellar dust clouds (Reddish, 1966). The locations of development through subsequent evolutionary phases may then be followed in the case of massive stars by observations of compact HII regions and dust-embedded OB stars, and in the case of stars of small mass by observations of T Tauri stars.

Three major difficulties are inherent in attempts to discover the relationship between the distributions of these objects and that of the interstellar gas; dark clouds appear to be the seat of star formation, yet globules can only be detected against a bright background; anomalous OH sources apparently require the presence of infrared protostars to stimulate them into emission; and the interstellar dust appears to be so dense in regions where stars form that the visible radiation cannot penetrate. The intercomparisons which follow are subject to these limitations.

2.0321 Young stars and globules Bok (1948) suggested that the globules are protostars; on the other hand van den Bergh (1967) and Herbig (1970) have argued that the globules are detached pieces of "elephant trunk" and suchlike structures and disappear through evaporation rather than by contraction to form stars. On the basis of either argument we may therefore expect to find fewer globules in older star groups than in newly formed ones.

Uncertainty arises in part from our lack of knowledge of the total masses of globules; the observational data give their angular diameters, and opacities (usually lower limits) due to dust grains, and it is not yet known if they contain gas as well as dust.

In the circumstances the possibility that the globules are protostars will be left for later examination, and only their distribution in relation to young stars will be considered.

A search for globules in 57 out of 66 OB clusters and associations previously examined for dust-embedded stars (Reddish, 1967c) has been carried out by Sim (1968b) and the numbers of globules found are given in Table 2.01.

TABLE 2.01

Numbers of Globules

	Certain	Probable	Doubtful
13 OB associations	44	42	32
44 OB clusters	24	23	19
Totals	68	65	51

If the clusters and associations searched for dust-embedded stars are arranged in four classes a to d (Reddish, loc. cit.) where in a systems all the stars have circumstellar obscuration and in d systems none have, and we select only those systems with good to moderate background brightnesses, then the numbers of certain and probable globules (giving half weight to the probables) are found to decrease along this sequence, Table 2.02.

TABLE 2.02

Systems with Good to Moderate Background Brightnesses

Class	Dust-embedded Stars	No. of Systems	No. of Globules	Average No. of Globules/ System
a	all	5	54	11
b	many	2	8	4
c	few	4	3	0.7
d	none	9	13	1.4

This is a sequence of age increasing from a few hundred thousand to several million years. The relatively larger number of globules in the younger systems indicates either that the globules have condensed into stars in the older systems, or that they have been evaporated. In the latter case it may be expected that the numbers of globules seen against bright nebulae would be related to the size or structure of the nebulae, but the survey by Roshkovsky (1954) shows that there is no such relation. The evidence therefore indicates that the globules are protostars, but it is not conclusive, and further evidence against is presented in the next section.

Herbig has pointed out (1962) that although T Tauri stars are always found in or near dark clouds, none has been found in a globule. However, the nature of a globule probably changes if and when a star forms from its core, and it may be that the anomalous OH emission sources and some of the infrared sources represent the later stages in the process of the formation of stars in globules.

Model calculations by Larson (1972b) indicate an evolutionary sequence globules → infrared objects → T Tauri stars. Spectral line emission by molecules has been detected coming from some globules (Bok and McCarthy, 1974).

2.0322 Anomalous OH emission sources If the anomalous OH 18 cm line emission sources are a stage in the contraction of a globule towards stellar density, a spatial correlation would be expected between the two kinds of objects. About half the OB clusters containing dust-embedded stars and globules were therefore searched for OH sources by Turner (1969a), but none were found. Anomalous OH emission sources were not in fact found associated with any of the young clusters containing dust-embedded stars, with globules, with Wolf-Rayet stars, or with most HII regions. Where the sources occur they generally appear to be associated with compact HII regions, with supernova remnants, or with infrared objects, but it is by no means certain that they always show such an association.

Turner (1969b) divided the OH emission sources into three classes:

(I) emission in the central pair of lines,

(II) emission in one or both of the satellite lines,

(III) absorption in all lines.

Class I sources appear to occur primarily at the edges of compact HII regions (Mezger and Höglund, 1967; Schraml and Mezger, 1969). Class IIa sources (1612 MHZ line strongest) are associated with infrared objects (Wilson and Barrett, 1968) but only a minority of infrared objects show OH emission (ibid; Eliasson and Groth, 1968). Class IIb sources (1720 MHZ line strongest) are associated with regions of non-thermal radio emission, possibly supernova remnants (Ball and Staelin, 1968), but not all OH sources fall neatly into these subdivisions (Booth, 1969).

Litvak (1969) has shown that the maser emission of OH may be pumped by infrared radiation, the satellite lines emitting in cool OH clouds and the central lines in hotter and turbulent ones. This appears to fit the association of satellite line emitters with known infrared sources and the central line emitters with compact HII regions but it remains to be shown that all anomalous OH emission sources are located in or close to infrared sources of sufficient intensity.

2.0323 Infrared sources Most infrared sources appear to be evolved late type stars such as Mira variables (Wing, Spinrad and Kuhi, 1967), probably further reddened by interstellar or circumstellar grains (Reddish, 1966). In the absence of a spectroscopic criterion for distinguishing these from possible protostars with any certainty, the latter have been selected by their association with other very young objects such as HII regions (Wynn-Williams and Becklin, 1974). In a survey of the Galactic plane at a wavelength of 100 μm, Hoffmann, Frederick and Emery (1971) catalogue 72 objects of which 60 are identified with radio sources or with bright or dark nebulae. The infrared sources in Orion are examples (Mezger, 1969) which are associated with HII regions and anomalous OH emission sources. The infrared objects in W3 appear to be compact HII regions having dust radiating at 150 K and producing 50 visual magnitudes of obscuration over and above the obscuration by dust surrounding the regions (Wynn-Williams, Becklin and Neugebauer, 1972). On the other hand, the four infrared sources found by Wilson and Barrett (1968) to be associated with anomalous OH emission sources do not lie in HII regions (Churchwell, Felli and Mezger, 1969). T Tauri stars emit an excess of infrared radiation (Mendoza, 1968) and it may be that the infrared sources associated with anomalous OH emission but not with HII regions are young stars of low mass such as T Tauri stars, or protostars which have not yet reached a high enough temperature to ionise the surrounding hydrogen. There is an infrared source radiating strongly at 350 μm in the Ophiuchus dark cloud (Simon et al, 1973) where there is evidence of abundant H_2 molecules and very recent star formation, and a cluster of infrared sources in a molecular cloud in Orion (Gatley et al, 1974).

Maps of millimetre wavelength radiation by molecules in regions such as W75, W3, NGC 2024 and NGC 2264 which contain infrared sources show large and extensive concentrations of molecules whether or not there is a compact HII region. Molecular cloud masses are generally in the range 10^4 - 3×10^5 M☉. Analysis of the observational data leads to the conclusion that these regions are aggregates of clouds of molecules in which star formation extends over a time $\geqslant 10^4$ yr, beginning with a dense dark cloud of molecules, forming protostars, infrared sources, OB stars and compact HII regions which subsequently expand (Morris et al, 1974).

2.0324 Compact HII regions A significant feature of the compact HII regions is the very high obscuration by interstellar grains of their central regions, especially the exciting stars. From a comparison of optical with radio data, Wynn-Williams (1971) deduces that obscuration exceeding 15 visual magnitudes results from dust concentrated around the compact regions in diameters 0.1 to 1.0 pc. Masses of HII are typically 20 M☉ per condensation with densities $n_{H+} = n_e \approx 10^4$ cm^{-3}. If the ratio of dust to gas is 'normal', the surrounding HI amounts to $\geqslant 40$ M☉. They appear to occur in regions of high gas density and the major concentration of HII regions between 4 kpc and 6 kpc from the centre of the Galaxy seems to be associated with regions of very high interstellar obscuration (section 2.031). Surveys indicate that 28% of all HII regions contain compact structures (Felli, Tofani and D'Adderio, 1974). It has been estimated that each compact HII region contains an aggregate of several O-stars. Evolutionary processes to be considered later lead to the view that compact HII regions are an early stage in the evolution of an O-association or cluster, and it follows that aggregates of newly formed O-stars are to be found embedded in dense gas clouds showing a very high obscuration by interstellar grains.

The spatial relationships of compact HII regions with infrared sources, anomalous OH emission sources, T Tauri stars, and dust-embedded OB stars, have been investigated by Churchwell, Felli and Mezger (1969). Except for the OH sources few of the objects have compact HII regions associated with them.

2.04 Summary

Many stars, and possibly all of them, form in aggregates. In spiral galaxies these aggregates and the dark interstellar clouds are gathered together into spiral arms. They form the most flattened subsystem in the Galaxy, indicating the lowest velocity dispersion.

The stars form in the dense, dark clouds. The gas in such clouds appears to be in molecular form, and obscuration by the high density of interstellar grains in them is so great that the most newly formed stars probably cannot be observed except in the infrared.

Bok globules and anomalous OH emission sources may be protostars; infrared objects, compact HII regions and dust-embedded OB stars may represent successive stages in the formation and evolution of massive stars, with T Tauri stars being newly formed stars of low mass. All of these various objects are located in or near dark clouds.

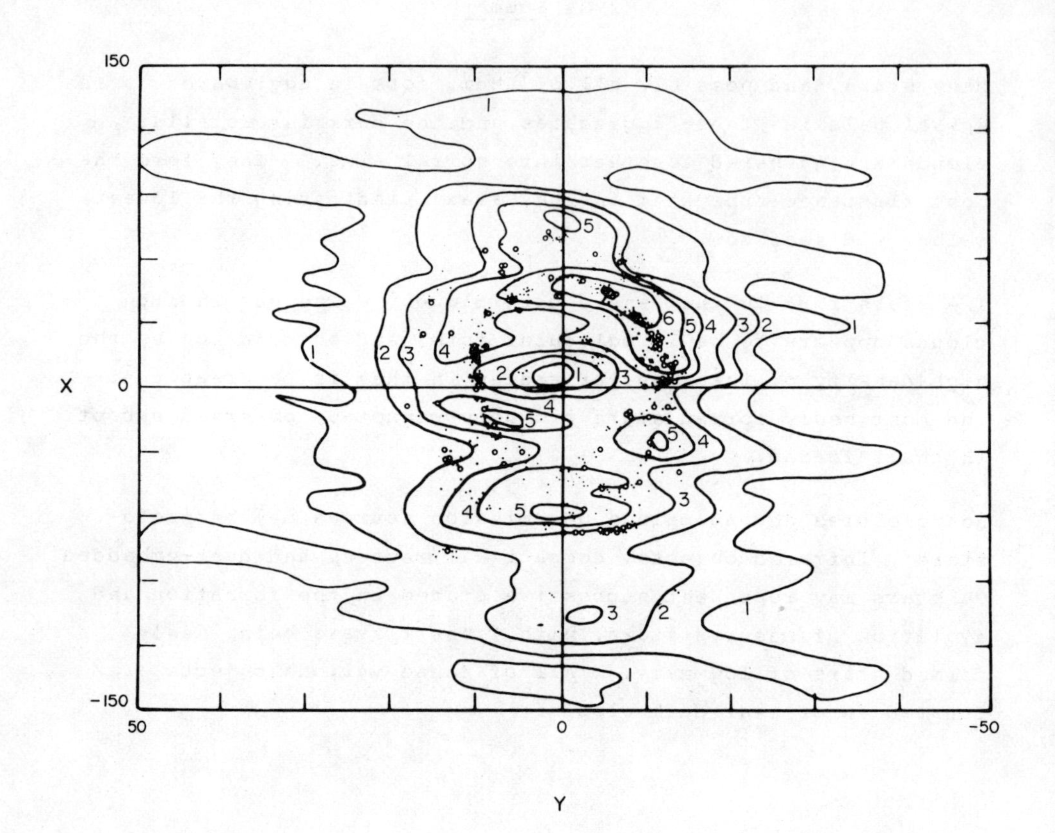

Fig. 2.04a
The distribution of 688 HII regions in M31, compared with that of the neutral hydrogen. The brightest HII regions are denoted by open circles.

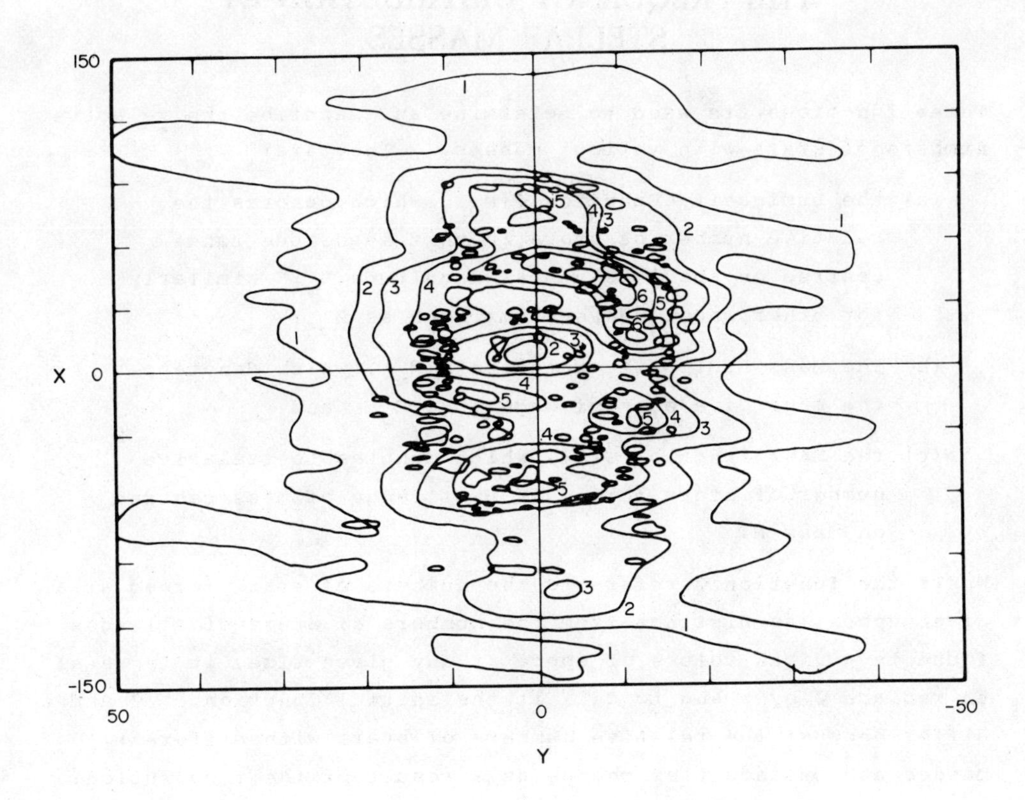

Fig. 2.04b
The distribution of 188 OB associations in M31 compared with that of the neutral hydrogen.

Chapter 3.

THE FREQUENCY DISTRIBUTION OF STELLAR MASSES.

Three functions are used to determine and describe the relative numbers of stars with various masses. They are:

(a) the Luminosity Function $\emptyset(M_v)$, which denotes the relative number of stars in unit magnitude range centred on absolute visual magnitude M_v; similarly for other magnitude systems such as M_{pg};

(b) the Mass Luminosity relation $M(M_v)$, which denotes the mass of a star of magnitude M_v; and

(c) the Mass Function $\psi(M)$, which denotes the relative number of stars formed in unit range of mass centred on mass M.

Where the function $\emptyset$ refers to the numbers of stars formed at a given epoch, as distinct from the numbers of stars of all ages found in a given volume of space at any given time, it is usual to replace $\emptyset$ by ψ and to call it the initial function. $\emptyset$ and ψ differ because the relative numbers of stars with different masses and luminosities change as a result of their evolution.

It may be expected that the relative numbers of stars formed with various masses will depend on the processes of star formation which, in turn, may be expected to depend on the physical conditions, chemical composition and state of motion of the material from which the stars form. Therefore in addition to finding the average form of $\psi(M)$, we need to know whether or not the form changes, and if so what causes it to change.

3.01 The Luminosity Function $\emptyset(M_v)$

Extensive investigations of the luminosity function in the general field of stars in the solar region of the Galaxy have been made by McCuskey (1966). The mean values he derived for visual magnitudes are reproduced in Table 3.01.

Within several hundred parsecs of the Sun, the luminosity function shows a considerable uniformity. The determinations of the luminosity function by McCuskey have been made in twelve regions distributed around the Galactic plane from $l^{II} = 44^{o}$ to $l^{II} = 214^{o}$, and the range of variation with differences in both distance and longitude is indicated by the vertical lines on Fig. 3.01.

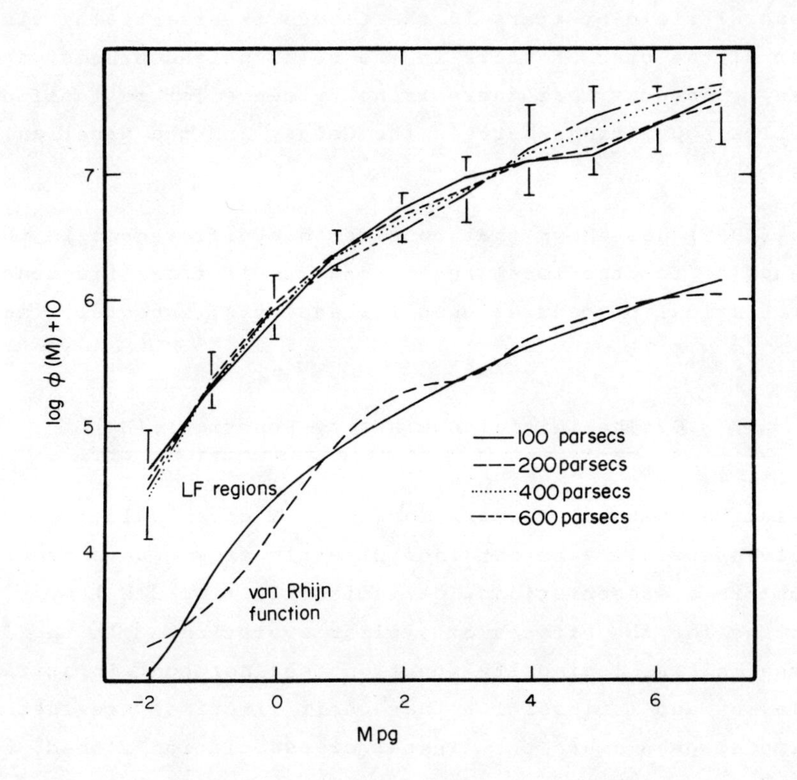

Fig. 3.01
Variations in the luminosity function. The upper curves show the luminosity function averaged over all LF regions at each of the distances 100, 200, 400 and 600 parsecs from the Sun. The range of variation about the means is indicated by vertical lines. The lower curves show the function averaged over all LF regions at all distances, and the van Rhijn function.

The total range of variation about the mean is by a factor $\pm$ 1.9 in the number of stars per unit magnitude range, averaged over all magnitudes. This may be compared to the total range of the mean luminosity function which is by a factor ~ 10^5:1.

Variations from the mean of much greater amount are found in the Magellanic Clouds (McCuskey loc. cit.), but it is probable that this is due to the data covering a wider range of types of star groupings, from old population II systems to young clusters and associations, from a general admixture of stars to groups of stars formed at a common epoch. The luminosity function for the general field of stars in the Clouds is essentially similar to that of the disc of stars in the solar neighbourhood, and McCuskey concludes that there is no evidence for real differences in stellar populations between the Galaxy and the Magellanic Clouds.

Warner (1972) has shown that considerable differences in the derived $\phi(M)$ for the local region can result from differences in the statistical procedures used for analysis, especially in the range $12 \leqslant M \leqslant 18$.

3.02 The Initial Luminosity Function $\psi(M_v)$

The relative numbers of stars formed in a given volume of space at a given epoch may be obtained directly from counts of stars in clusters or associations, or indirectly from $\phi(M_v)$ by correcting for the effects of stellar evolution. In the first case the initial luminosity function does not contain stars of such masses and luminosities that their lifetimes are less than the time elapsed since the cluster or association formed, nor stars of masses which take longer than that time to evolve to a visible state. In the second case the initial luminosity function is an average over time, weighted according to the dependence of the rate of star formation on time, from the earliest time at which surviving stars formed. Uncertainties in the $\psi(M_v)$ derived in this way result from our lack of knowledge both of the rate of star formation at different epochs in the volume of space to which the data for $\phi(M_V)$ refers, and of the interchange of stars between that volume and its surroundings.

Salpeter (1955) was the first to compute the initial luminosity function $\psi(M_V)$ from $\phi(M_V)$ with the assumptions (i) that stars evolve off the main sequence when ten percent of their hydrogen has been converted to helium, and (ii) that the rate of star formation in the local region of the Galaxy has been constant over the last 5 x 10^9 yr.

TABLE 3.01

Luminosity Function (Solar Neighbourhood)

Column 3 gives the Salpeter Initial Luminosity Function re-derived from McCuskey's data in Column 2. Column 4 gives data for Cyg OB2 and the blue globular clusters in the LMC. Column 5 gives Wood's data for E and SO galaxies.

(1)	(2)	(3)	(4)	(5)
M_V	log $\phi(M_V)$	log $\psi(M_V)$	log $\psi(M_V)$	log $\psi(M_V)$
-8	-	-	4.67	-
-7	-	-	4.92	-
-6	2.23	-	5.22	-
-5	2.97	-	5.58	-
-4	3.60	5.83	5.95	-
-3	4.10	6.26	6.15	-
-2	4.80	6.63	-	-
-1	5.54	6.92	-	-
0	6.16	7.06	-	-
1	6.60	7.15	-	-
2	6.83	7.18	-	-
3	7.07	7.10	-	-
4	7.25	7.23	-	-
5	7.56	7.56	-	7.56
6	7.66	7.66	-	8.14
7	7.48	7.48	-	8.56
8	7.58	7.58	-	8.86
9	7.72	7.72	-	9.19
10	7.88	7.88	-	9.42
11	7.97	7.97	-	9.58

TABLE 3.01 Cont.

(1)	(2) log $\phi(M_v)$	(3) log $\psi(M_v)$	(4) log $\psi(M_v)$	(5) log $\psi(M_v)$
12	8.00	8.00	-	9.71
13	8.05	8.05	-	-
14	8.05	8.05	-	-
15	8.02	8.02	-	-
16	7.96	7.96	-	-

Repeating the calculations made by Salpeter but using McCuskey's $\phi(M_v)$, the initial luminosity function $\psi(M_v)$ given in column 3 of Table 3.01 results.

The $\psi(M_v)$ derived directly from young open star clusters show considerable differences for stars fainter than the Sun (van den Bergh and Sher, 1960), but for stars brighter than this the average over the clusters agrees well with that derived from $\phi(M_v)$ (McCuskey, 1966).

It was pointed out early in this section that in addition to knowing the average $\psi(M_v)$ we also desire to know if there are real differences from the average, dependent on local conditions in the matter from which stars form. The difficulty, of course, is that when the stars have formed, so that we can determine their luminosity functions, the conditions in the matter from which they formed have been changed. We can search for any dependence of $\psi(M_v)$ on such things as the total mass, the size, and the age of the system, and the chemical compositions, the random velocities and the rotations of the stars; but we have no evidence of the temperature and state of combination of the gas at the time the stars formed from it.

An analysis by Lindoff (1967) of 177 open clusters indicates that the gradient of the initial luminosity function has not changed with time during the last 6×10^8 yr. Evidence for much older systems is lacking; luminosity functions for globular clusters (Simoda and Kimura, 1968) do not extend to magnitudes much fainter than those affected by evolutionary processes.

Young open clusters generally have small total masses in comparison with the globular clusters, so that the absence of information for old clusters also limits attempts to discover any dependence on total mass. There are, however, a few massive young star clusters - the association Cygnus OB2 (Reddish, Lawrence and Pratt, 1966) and the blue globular clusters in the Magellanic Clouds (Woolley, 1960).

Cygnus OB2 is estimated to have a total mass about 3×10^4 $M_\odot$ (Reddish et al, loc cit). The luminosity function does not differ significantly from that derived by Salpeter for stars with luminosities $-4 \leqslant M_V \leqslant +2$, but there may be a larger number of brighter stars in the association. The data does not extend to stars fainter than $M_V = +2$.

The blue globular clusters also show more stars brighter than $M_V = -4$ in comparison with the Salpeter Initial Luminosity Function.

The data on Cygnus OB2 and the blue globular clusters in the LMC have therefore been combined to strengthen at the bright end the data given in Table 3.01.

Dixon (1970) concludes that the average $\emptyset(M)$ for stars formed during the last 5×10^9 yr continues to rise beyond $M_V = +6$ whereas for stars in recently formed clusters it declines. It may be noted in passing that this is in agreement with the theoretical expectations shown on Fig. 13.05.

It is difficult to discover any dependence of the initial luminosity function on the chemical composition of the material out of which the stars formed, because those clusters which differ considerably in the chemical compositions of their stars are all old and hence have only faint stars remaining on their main sequences. For this reason nothing is known of the effect of chemical composition on the initial luminosity function.

Evidence regarding the effect of rotational velocity is meagre. For the young cluster M39 the rotational velocities of the fainter stars are high in comparison with, for example, the Pleiades, and the number of stars falls rapidly at these faint luminosities (Reddish, 1965). It is not clear whether the formation of these faint stars has been inhibited or just delayed, and more systematic data of this kind are required before firm conclusions can be drawn.

In summary, the initial luminosity function does not appear to vary significantly with position in the Galaxy over distances of hundreds of parsecs; it does not depend on the total mass of the system. There is meagre evidence of a possible dependence on stellar rotation, and no significant data regarding possible effects of differences in age or chemical composition.

These considerations apply to the gradient of the luminosity function as a function of magnitude, i.e. $\frac{d\psi(M_v)}{d\,M_v}(M_v)$, between lower and upper limits of luminosity. We need to enquire also into possible changes in these limits.

3.021 The Brightest and Faintest Stars Formed

The initial luminosity function in Table 3.01 extends down to $M_v = 16$. It is derived from data which includes that given by Luyten (1968), who found the stars within 10 parsec of the Sun from a proper motion survey. That sample is expected to be seriously incomplete at $M_{pg} = 20$ and may be partially incomplete at $M_{pg} = 17$. It shows a maximum in the luminosity function at $M_{pg} = 15.7$; Luyten considers that this is probably real, but it is so near to the magnitude at which incompleteness is expected to begin that the reality of the decrease in the initial luminosity function for stars fainter than $M_{pg} = 16$ must be regarded as doubtful.

Analysis of the integrated spectra of other galaxies by Wood (1966) shows that dwarf M stars occur in them in great abundance. Relative to the numbers at $M_V = +5$, there are fifty times as many stars at $M_V = +12$ as in the solar neighbourhood. The data for the range $5 \leqslant M_V \leqslant 12$ for the average of a number of E and SO galaxies has been obtained from Wood's Fig. 5 and is given in column 5 of Table 3.01. We do not know whether this difference represents an excess of dwarf stars in E and SO galaxies or a deficiency in the solar neighbourhood, or to what extent it may be due to selection effects and incompleteness in our local data. Wood's luminosity function is still rising towards fainter stars at $M_V = 12$, and he points out that his integrated galactic models are still deficient in mass, a deficiency which may be made good by the addition of very large numbers of much fainter stars.

Other evidence relating to the existence of stars of very low mass and luminosity will be given in section 3.04.

The theory of stellar structure indicates an upper limit to the mass of a stable star on account of vibrational instability when most of the internal pressure is due to radiation (Schwarzschild and Harm, 1959; Larson and Starrfield, 1971). Such a star will be expected to have a mass $\leqslant 100 \pm 40$ M⊙ and an absolute visual magnitude of about $M_V = -7$. No binary star system has yet been found for which masses have been derived which conflict with the evidence of the theory of stellar structure. This does not mean that more massive stars may not attempt to form, and are dispersed or explode. Appenzeller (1970) has shown that large mass stars (e.g. 130 M⊙) are not vibrationally unstable on a short time scale but lose mass over a period $\sim 10^6$ yr. It is difficult to apply an observational test to this conclusion because, as has been pointed out earlier, the formation of stellar masses from a gas cloud of finite mass results in a statistical upper limit to the most massive star formed.

For example, in the case of the association Cyg OB2 which has a total mass $\sim 3 \times 10^4$ $M_\odot$, the application of the initial mass function to be discussed in section 3.04 leads to an expected largest stellar mass of 240 $M_{\odot}$; extrapolation of the observed initial mass-luminosity relation to this mass leads to an expected brightest star at $M_v = -9$, which is close to that observed (Reddish, 1967b; 1967c). This is probably the brightest known star in our Galaxy but the uncertainties due to probable errors in distance and obscuration are considerable.

It is also understood that supermassive objects with masses of order 10^8 $M_{\odot}$ may form.

Since we are concerned in this treatise with the formation of stars rather than with their subsequent evolution, we must be concerned with the frequency distribution of those fragments of interstellar clouds which have enough gravitational potential to contract to form objects which, in a very broad sense, we may term stars regardless of whether or not such objects are stable.

We must bear in mind the possibility that the upper and lower limits of the initial luminosity function may change with time, as a result of changes in physical conditions. On account of the extreme faintness of stars at the faint end of the luminosity function, and the short lifetimes of those at the bright end, it is unlikely that direct observations will provide much useful relevant information, and we must have recourse to indirect evidence.

There are strong arguments derived from the rate of growth of element abundances in the Galaxy (Reddish, 1968a) that stars of much greater mass than are forming now were formed in the early history of the Galaxy, and differences in the mass-luminosity ratios of galaxies may be due to differences in the numbers of faint dwarf stars they contain (Reddish, 1961; 1968c; 1969a,b). These conclusions depend, however, on interpretations involving an understanding of the evolution of galaxies especially in relation to the processes of star formation in them, and they are more relevant to the theoretical considerations in Part 2, where they are discussed in more detail.

3.03 The Initial Mass-Luminosity Relation

As a star contracts to the main sequence its central temperature rises according to the relation $T_c \propto \mu M/R$. When the temperature has become high enough for nuclear reactions to synthesise helium from hydrogen, the molecular weight μ increases slowly at almost constant T_c, and the radius R increases correspondingly. Consequently the initial main sequence of stars with various masses is a locus of points of minimum radius.

Stellar masses, being derived from orbital data of detached binary stars not selected according to any criterion of age, will include some stars which have not yet contracted to the initial (zero age) main sequence and others which are evolving away from it. When their radii are plotted against their masses the distribution of points should be sharply bounded on the side of low radius, and have a dispersion on the side of larger radius due to contracting and evolved stars. Such a distribution is indeed observed (Reddish, 1962a), Fig. 3.02. Only those points which are included within a normal error dispersion about a locus of minimum radius represent stars on the initial main sequence, and the relation between their masses and absolute visual magnitudes is given on Fig. 3.03. Stars selected in this manner are found to delineate a main sequence on the colour-magnitude diagram in quite good agreement with the initial main sequence defined by non-binary stars in open clusters (Reddish, loc cit).

More recent investigations of the mass-luminosity relation by Eggen (1967, 1969) have given a result not differing significantly from Fig. 3.03 for unevolved main sequence stars in the UMa cluster and Sirius group, but masses lower by about one third at a given magnitude are found for stars in the Hyades and for other stars which do not show an ultraviolet excess in their colours. If the ultraviolet excess is interpreted in the usual manner as indicating a relative deficiency of metals in the stellar atmosphere, the implication is that the initial mass-

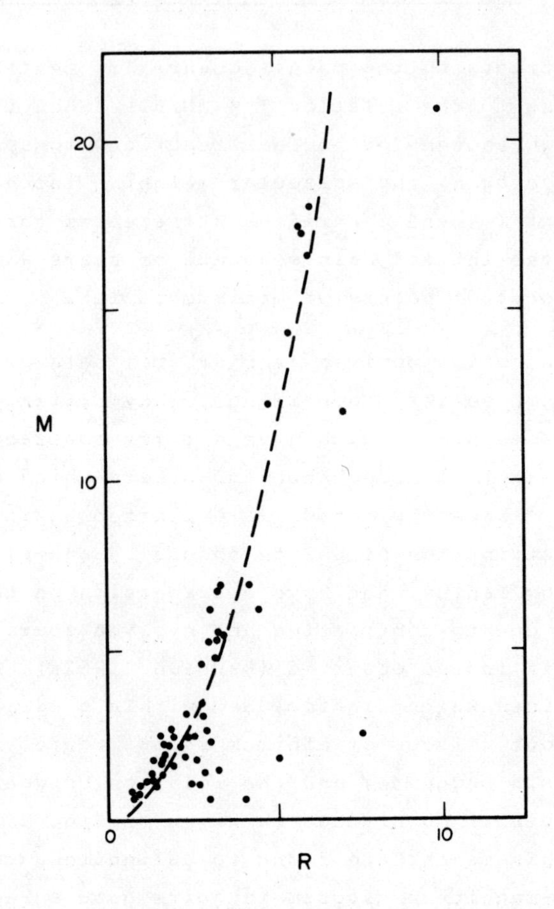

Fig. 3.02

Mass-radius relation for detached binary systems. Those to the left of the broken line are unevolved main sequence stars.

luminosity relation depends on the chemical composition of stars. It may therefore vary with both time and position in the Galaxy. The quantity of data available is, however, too small to enable any dependence of the empirical mass-luminosity relation on composition to be set down in any detail or with any certainty, and the low dispersion on Fig. 3.03 indicates that it defines a

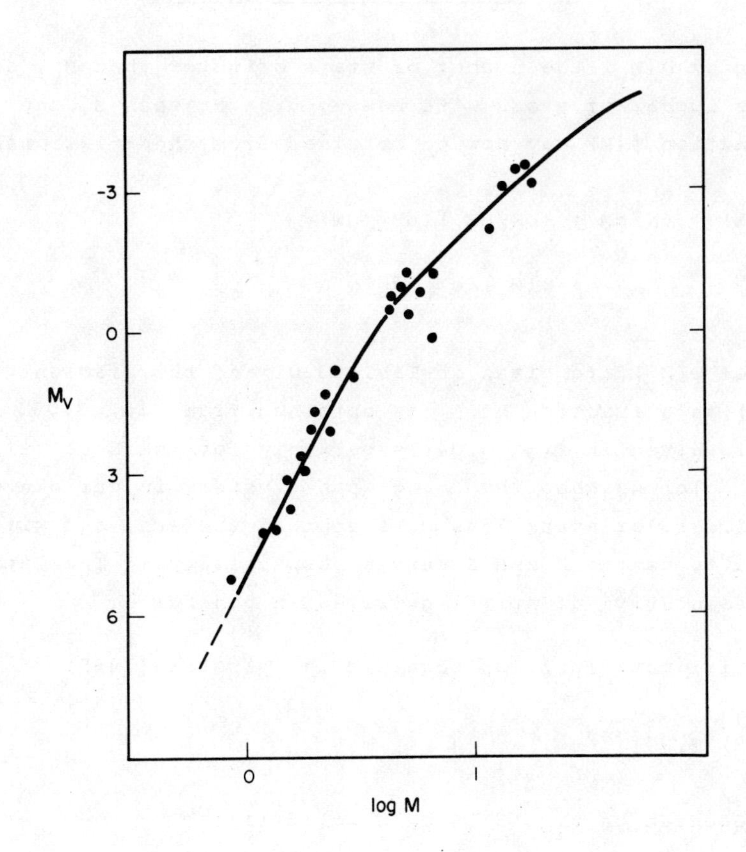

Fig. 3.03
Mass-luminosity relation for unevolved main sequence stars.

relation satisfied by the great majority of unevolved binary stars at the present time. It will therefore be used in the next section to convert the initial luminosity function to an initial mass function.

3.04 The Initial Mass Function

Denoting by $N(M_V)$ the number of stars brighter than M_V, and by $N(M)$ the number of stars with masses greater than M, the initial mass function $\psi(M)$ may now be obtained from the relationships

$$\psi(M) = dN/dM = (dN/dM_V)(dM_V/dM)$$

$$= \psi(M_V)(dM_V/dM) \quad (3.01)$$

Values of $\psi(M_V)$ are given in Table 3.01 and the gradient (dM_V/dM) as a function of M_V is obtained from Fig. 3.03. The result is given in Fig. 3.04, separately for the E galaxies, and for the solar neighbourhood and open clusters in our own Galaxy. These differ for stars less massive than the sun, and will be referred to as the E and S curves respectively. The latter may be representative of spiral galaxies in general.

The initial mass function obtained by Salpeter (1955)

$$\psi(M) \propto M^{-2 \cdot 35} \quad (3.02)$$

is represented on Fig. 3.04 by a solid line.

The small scale bends in the curves are probably not significant, except for that at $\log M = -0.15$ on curve S. This is the mass of a star which has a main sequence lifetime about equal to the estimated age of the Galaxy. This part of the mass function is derived from field stars in the solar neighbourhood, on the basis of an assumed dependence of the rate of star formation on time, and with corrections applied for the lifetimes of more massive stars being less than the interval over which stars have been forming. A spurious kink at $\log M = -0.15$ could result from an error in these factors. For example, it may be that the rate of star formation in the Galaxy during the first 10^9 yr or so was less than has been assumed by a factor of two or three. If this were so the part of the S curve for $\log M \leqslant -0.15$ would be raised to the Salpeter line.

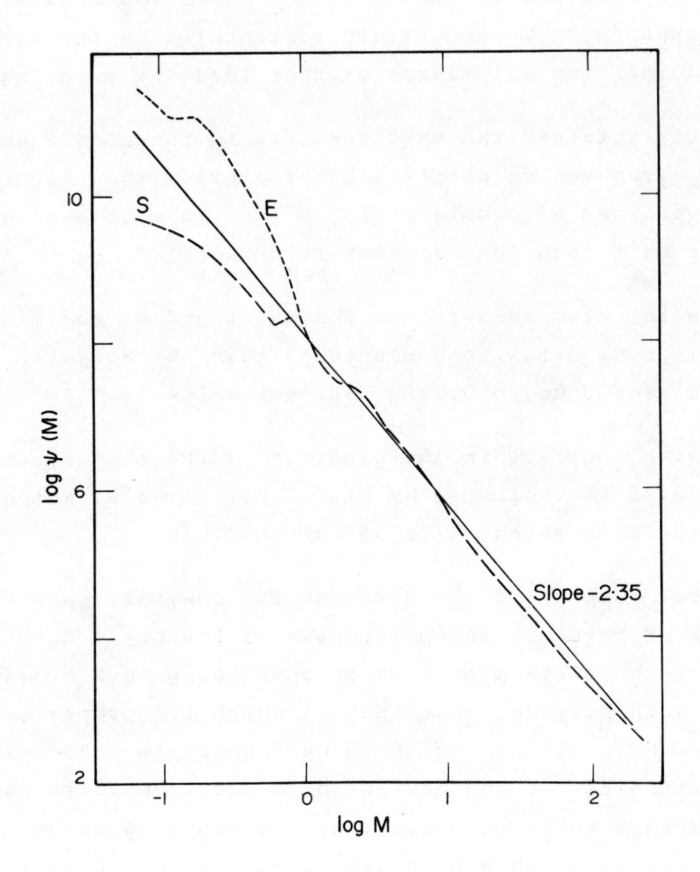

Fig. 3.04
Frequency distributions of the masses of stars formed in elliptical (E) and spiral (S) galaxies. $\psi(M)$ is the relative number of stars formed in unit range of mass at mass M.

At this same value of mass, log M = -0.15, the E curve shows a rapid rise. This similarly may be due to an exceptionally high rate of star formation in the early history of these galaxies, rather than to a real difference in the initial mass function.

The steepening of the initial mass function at high masses shown in the analysis by Warner (1961a) and by Salpeter's data is not confirmed by the later investigations using more extensive data on bright star counts (Reddish, 1962a, 1962c and Fig. 3.04). The function appears to be accurately represented by the Salpeter equation (3.02) for all masses greater than one solar mass.

Huang (1961) obtained the mass function in the mass range 0.4 $M_\odot$ to 0.05 $M_\odot$ from van de Kamp's list of stars within 5 pc of the Sun, and obtained a formula $F(M) \propto M^{-1 \cdot 35}$, a much less steep dependence on M than the Salpeter relation.

Warner (1961b) also re-examined the faint end of the mass function in some detail and concluded that the Salpeter relation held for masses down to 0.25 $M_\odot$ but not below.

A progressive decrease in the gradient of the mass function below M = 0.25 is confirmed by Fig. 3.04, curve S, which is based on the more recent data of Luyten (1968).

The physical reality of the decrease is, however, open to doubt on grounds of possible incompleteness of the basic data. Although the absolute magnitude corresponding to M = 0.25 $M_\odot$ is M_v = +11, much brighter than that at which the proper motion search by Luyten (1968) begins to be incomplete, the volume of space surrounding the Sun may not be a representative sample of the Galaxy as a whole on account of the changing volume within which star formation has been taking place, and the corrections applied for the time dependence of the rate of star formation in the local region may be inadequate.

On the other hand it may be, as suggested earlier, that the initial rate of star formation assumed in the analysis is too high for spiral galaxies and too low for ellipticals.

In any case, it is convenient to regard the Salpeter Law (equation 3.02) as a basic generalisation, bearing in mind the possibility of deviations from it for stars with masses $< 1\ M_\odot$.

3.041 The Largest and Smallest Stellar Masses

It was mentioned in section 3.021 that the theory of stellar structure may set an upper limit at about 60 $M_{\odot}$ to the mass of a stable star (Schwarzschild and Harm, 1959) but that two other considerations are relevant to the present study. Firstly, if cloud fragments having masses > 60 $M_{\odot}$ attempt to form stars, but fail on reaching the unstable condition, the frequency distribution of their masses is nevertheless important to us in studying the fragmentation of an interstellar cloud and the formation of protostars. Secondly, in any cloud of finite mass, there is an upper limit to the mass of a protostar fragment formed from it. If the masses of the protostar fragments follow the Salpeter distribution function, and if, as seems probable, the lower limit to the mass distribution is determined by physical conditions in the cloud, then the upper limit will be very much less than the total mass of the cloud and may be estimated as follows.

If N is the total number of stars formed in some system, with masses between upper and lower limits M_2 and M_1, the number in the mass range M to M_2 is

$$N \int_{M}^{M_2} F(M)\,dM \qquad (3.03)$$

since $\int_{M_1}^{M_2} F(M)\,dM = 1$ by definition. (3.04)

The numbers of stars formed in any range of mass decrease with increasing mass according to equation (3.02), and the largest stellar mass will be found within the mass range M to M_2 that contains only one star; in this case

$$N \int_{M}^{M_2} F(M)\,dM = 1 \tag{3.05}$$

If M_t is the total mass of the system and $\bar{M}$ is the average mass of a star formed in it, then

$$N = M_t/\bar{M} \tag{3.06}$$

Using (3.02) we have

$$\bar{M} = \int_{M_1}^{M_2} F(M)\ M dM \Big/ \int_{M_1}^{M_2} F(M)\,dM$$

$$= 3.9\ (M_1^{-0.35} - M_2^{-0.35})/(M_1^{-1.35} - M_2^{-1.35}), \tag{3.07}$$

or

$$\bar{M} \approx 3.9\ M_1,\ M_2 >> M_1 \tag{3.08}$$

Equations 3.02, 3.04, 3.05, 3.06 and 3.08 give

$$\int_{M}^{M_2} M^{-2.35}\ dM \Big/ \int_{M_1}^{M_2} M^{-2.35}\ dM = 3.9\ M_1/M_t. \tag{3.09}$$

Since M_2 can be made infinite, $M_2 >> M >> M_1$ and we have finally

$$M \approx 0.36\ M_1^{0.26}\ M_t^{0.74} \tag{3.10}$$

as the approximate upper limit to the mass of a star formed due to the finite total mass M_t of the protostar gas cloud.

The value to be assigned to M_1 is uncertain, and will be discussed below. It probably lies between 0.1 $M_\odot$ and 0.025 $M_\odot$, but equation (3.10) is insensitive to the precise value, and using $M_1 = 0.05\ M_\odot$ will lead to an error of not more than 20% in M. With this value for M_1, the mass M of the largest star formed is given in Table 3.02 for various total cloud masses M_t.

TABLE 3.02

Mass of Largest Star Formed as a Function of Cloud Mass

(Units of $M_\odot$)

Mass of Cloud	Mass of Largest Star
10^5	830
10^4	150
10^3	28
10^2	5

Evidently, statistical limitations alone would be sufficient to account for the absence of stars with masses greater than the stability limit of about 60 $M_\odot$ in all stellar aggregates with total masses $\leqslant$ 3000 $M_\odot$.

The association Cygnus OB2 has a total mass of about $3 \times 10^4\ M_\odot$ (Reddish, Lawrence and Pratt, 1966). Equation (3.10) gives, for the mass of the largest star formed, 240 $M_\odot$. Ignoring the stability problem for the moment, extrapolation of the mass-luminosity relationship in Fig. 3.03 gives an absolute visual magnitude $M_v \approx -9$ for this mass. Estimates of M_v for the brightest star in Cygnus OB2, Schulte No. 12, range from about -9 to -12 (Reddish, 1967c). For a star of mass 60 $M_\odot$, Fig. 3.03 gives $M_v \approx -6$. The indication of Cygnus OB2 is therefore that the largest star has a mass of ≈ 240 $M_\odot$ set by statistical limitations, not by stability criteria, an indication that merits further investigation. It would be of interest to know if Schulte No. 12 is a binary.

Any lower limit to the stellar mass function is more difficult to determine directly on account of the extreme faintness of the objects involved.

Several components with masses below 0.1 $M_{\odot}$ have been detected in binary systems; Ross 614B, 0.08 $M_{\odot}$, (Lippincott, 1955); L-726-8, with components 0.035 $M_{\odot}$ and 0.044 $M_{\odot}$ (Luyten, 1956); 61 Cygni, 0.008 $M_{\odot}$, (Strand, 1956); Lande 21185, 0.01 $M_{\odot}$, and the companion to Barnard's star, 0.0015 $M_{\odot}$, (van de Kamp, 1963). The data are evidently not adequate to give any indication of the frequency of occurrence of stars with such low masses.

An estimate of the lower limit to the initial mass function can be made by calculating the value necessary to satisfy the estimated mass density in the Galactic plane.

Let the faintest stars to which the initial mass function has been directly determined have masses M. Of the mass per unit volume contributed by stars, the fraction in the mass range from M to the lower stellar mass limit M_1 is

$$f = \int_{M_1}^{M} F(M)MdM \Big/ \int_{M_1}^{M_2} F(M)MdM \tag{3.11}$$

and using equation (3.02) this becomes, with some reduction

$$M_1 = M(1 - f)^{2.86} \tag{3.12}$$

The luminosity function in the solar neighbourhood determined by Luyten (1968) gives a mass density 0.064 $M_{\odot}$ pc^{-3} due to stars brighter than $M_{pg} \approx 18$, that is, stars with masses greater than 0.063 M_{O} (Allen, 1964; this also accords with a linear extrapolation of Fig. 3.03). The total mass density is estimated to be 0.11 $M_\odot$ pc^{-3} (Woolley and Stewart, 1967), leaving 0.046 $M_\odot$ to be accounted for by interstellar matter and by stars with masses less than 0.063 $M_\odot$.

The mass density of interstellar gas is uncertain. Oort (1958a) gives 0.025 $M_\odot$ pc^{-3} for HI, but suggests that, since some of the hydrogen probably occurs in clouds which are optically thick, the true amount may be nearer 0.037 $M_\odot$ pc^{-3}.

On the other hand recent measurements of the strength of interstellar Lyman α in the spectra of stars indicates that the amount of HI is lower by a factor ten than has been deduced from 21 cm line measures, that is, 0.003 $M_\odot$ pc^{-3}. Besides the HI, however, there may be a substantial mass density due to H_2, although this is unlikely to be very large in the solar region.

Depending upon which mass density of interstellar matter is adopted, 0.025, 0.037 or 0.003, the amount of mass remaining unaccounted is 0.021, 0.009 or 0.043, all in units of $M_\odot$ pc^{-3}. If this 'missing' mass is presumed to be in stars less massive than M = 0.063 $M_\odot$, then equation (3.12) gives the corresponding values of the lower limit to the initial mass function as M_1 = 0.034, 0.049 or 0.015 $M_\odot$ respectively.

Alternatively, large numbers of dwarf M stars brighter than M_v = 13, contributing 0.05 $M_\odot$ pc^{-3}, may provide the missing mass (Murray and Sanduleak, 1972) keeping M_1 at 0.063. Otherwise, in summary, the local mass density indicates a lower limit to the initial mass function,

$$M_1 = 0.033 \pm 0.010\ M_\odot \tag{3.13}$$

which is not significantly different from the value 0.027 similarly derived from earlier data (Reddish, 1962a). This value for M_1 is not inconsistent with the sparse data about low mass stars summarised above.

It is assumed in the calculation leading to equation (3.13) that the index -2.35 in the initial mass function, represented by the

Salpeter relation (3.02), is maintained unaltered until a sharp cut-off is reached at mass M_1. It may be that there is in fact a more gradual decline in the numbers of stars with small mass, in which case M_1 will represent an effective lower limit.

3.05 Summary

The frequency distribution of the masses of stars formed, the so-called initial mass function, is well represented by

$$F(M) \propto M^{-2.35}, \quad M_2 \leqslant M \leqslant M_1,$$

where F(M) is the number of stars per unit range of stellar mass,

$$\infty \geqslant M_2 \geqslant 60\ M_\odot,$$

and $M_1 = 0.029 \pm 0.013\ M_\odot$. (3.14)

The mass function shows a high degree of uniformity, although there are few data regarding any possible changes with time. It may be that E galaxies have relatively more stars formed with masses $M \leqslant 1\ M_\odot$, and S galaxies relatively fewer, but the effect which leads to this conclusion could be caused by the rate of star formation in E galaxies being relatively high in the first 10^9 years or so, and that in S galaxies relatively low.

Chapter 4.

THE FREQUENCY DISTRIBUTION OF THE MASSES OF AGGREGATES OF STARS.

If the process by which the whole of an aggregate of stars forms from interstellar clouds is the same as that which forms individual stars, it is to be expected that the mass spectrum of stellar aggregates will show some relationship to that of stars.

An investigation of the mass spectrum of stellar aggregates may be made using data on star clusters and associations in the Galaxy.

In principle, we require, for every open cluster and OB association, the number of stars per unit range of stellar mass; in each case this would be integrated over all stellar masses to give the total mass of the system.

There are three simplifications and one complication.

Firstly, given the total mass of one of the aggregates we require only the <u>relative</u> total masses of the aggregates.

Secondly, since the dependence on stellar mass of the relative number of stars formed per unit range of stellar mass in an aggregate shows a considerable uniformity and probably follows a fundamental law (the initial stellar mass distribution function), and we will assume this to be true, we only require the number of stars formed in a given mass range in each aggregate, in order to obtain the relative total masses of the aggregates.

Thirdly, because there is a reasonably universal relationship between stellar mass and stellar luminosity, we require only the number of stars formed in a given range of absolute magnitude to obtain the relative total masses of the aggregates.

The complication is that, on account of evolutionary effects, the number of stars observed now in each magnitude range is not the same as at the time of formation, the discrepancy increasing towards the brighter stars (which may now have disappeared). Therefore it is necessary to reproduce the initial luminosity function (the number of stars formed in each unit magnitude range) for each aggregate by fitting the observed luminosity function of each aggregate to the shape of the standard initial luminosity function $\psi(M_V)$. The intercepts with the ordinate will then give the relative total numbers of stars, and hence the relative total masses, of the aggregates at their time of formation.

This method has been applied to seventy-two open clusters and twelve OB associations and involved counting the number of stars in unit range of absolute magnitude for each aggregate.

With the exception of three associations, the absolute magnitude of any star in an aggregate was found by the formula:

$$m = M + 5 \log_{10} r - 5 + 3E_{B-V}$$

where E_{B-V} = average colour excess for the whole aggregate.

For three associations, Cassiopeia III, Cassiopeia IV and Gemini I, the colour excess for each star in the association was taken and, for one other association, Cepheus IV, both methods were used.

After the number of stars per unit range of absolute magnitude had been counted for an aggregate, a graph of $\log_{10}$ (number of stars in unit range of absolute magnitude) against the absolute magnitude was fitted to the shape of the standard initial luminosity function by placing it over the standard initial luminosity function, matching the abscissae before moving it vertically until the points of the observed curve best matched the standard initial luminosity function. The intercept with the ordinate gave the relative total mass.

The results for Cepheus IV show that, in this case at least, the relative masses obtained by using two alternative methods for finding the absolute magnitudes of the stars were different, being 24.7 when one value of E_{B-V} was used for the whole association and 9.5 when E_{B-V} was determined for each star. As only one association was done by both methods, no general rule may be inferred. The values of the other three OB associations, where E_{B-V} was determined for each star, Cassiopeia III, Cassiopeia IV and Gemini I, must, however be treated cautiously as the calculated relative masses may be underestimated.

The majority of the data for open clusters were taken from the same sources (Hoag _et al_, 1961; Becker, 1963; Lindoff, 1967) and hence the counting procedure may be assumed to be consistent. The counts for the OB associations, however, were taken from various sources and there are probably inconsistencies caused by different astronomers' criteria. For fainter absolute magnitudes, the apparent magnitudes approach the plate limit and this introduces uncertainties in the number of stars per unit range of absolute magnitude counted.

TABLE 4.01

THE TOTAL MASSES OF THE AGGREGATES RELATIVE TO THE MASS OF TRUMPLER 1

Aggregate	$\log_{10}$ (Relative total mass)	Relative total mass
NGC 129	0.872	7.4
NGC 225	0.294	2.0
NGC 457	0.964	9.2
NGC 581	0.711	5.1
NGC 654	0.872	7.4
NGC 663	1.097	12.5
NGC 744	0.729	5.4

Aggregate	$\log_{10}$ (Relative total mass)	Relative total mass
NGC 957	0.910	8.1
NGC 1027	0.455	2.9
NGC 1245	0.835	6.8
NGC 1242	0.406	2.5
NGC 1444	0.617	4.1
NGC 1502	0.600	4.0
NGC 1528	0.728	5.3
NGC 1545	0.455	2.9
NGC 1647	0.420	2.6
NGC 1662	0.309	2.0
NGC 1664	0.722	5.3
NGC 1778	0.430	2.7
NGC 1893	1.052	11.3
NGC 1907	0.787	6.1
NGC 1912	0.878	7.6
NGC 2099	0.921	8.3
NGC 2129	0.567	3.7
NGC 2168	0.902	8.0
NGC 2169	0.110	1.3
NGC 2251	0.508	3.2
NGC 2264	0.289	1.9
NGC 2287	0.663	4.6
NGC 2301	0.392	2.5
NGC 2323	0.901	8.0
NGC 2324	0.826	6.7
NGC 2353	0.672	4.7
NGC 2422	0.411	2.6
NGC 6475	0.313	2.1
NGC 6494	0.789	6.2
NGC 6531	0.506	3.2
NGC 6611	1.054	11.3
NGC 6694	0.556	3.6
NGC 6709	0.575	3.8

Aggregate	$\log_{10}$ (Relative total mass)	Relative total mass
NGC 6755	1.163	14.6
NGC 6802	0.890	7.8
NGC 6823	1.004	10.1
NGC 6830	0.706	5.1
NGC 6834	0.766	5.8
NGC 6866	0.751	5.6
NGC 6871	0.787	6.1
NGC 6910	0.908	8.1
NGC 6913	0.714	5.2
NGC 6940	0.599	4.0
NGC 7031	0.610	4.1
NGC 7062	0.724	5.3
NGC 7063	0.339	2.2
NGC 7067	0.933	8.6
NGC 7086	0.415	2.6
NGC 7128	0.763	5.8
NGC 7142	0.468	2.9
NGC 7160	0.620	4.2
NGC 7209	0.607	4.0
NGC 7235	1.080	12.0
NGC 7261	0.466	2.9
NGC 7380	0.861	7.3
NGC 7510	1.373	23.6
NGC 7654	0.911	8.1
NGC 7790	0.296	2.0
IC 1805	1.012	10.3
IC 1848	0.642	4.4
IC 4996	0.899	7.9
Trumpler 1	0.000	1.0
Trumpler 2	0.469	2.9
Trumpler 35	0.945	8.8
Hyades	0.503	3.2
Cassiopeia V	1.723	52.8

Aggregate	$\log_{10}$ (Relative total mass)	Relative total mass
Cepheus III	1.278	19.0
Lacerta I	1.170	14.8
Orion I	1.482	30.3
Perseus II	0.651	4.5
Sagittarius V	2.129	134.6
Cepheus IV (i)	1.392	24.7
Cepheus IV (ii)	0.978	9.5
Cassiopeia III	0.130	1.3
Cassiopeia IV	0.569	3.7
Gemini I	1.646	44.3
Cygnus II	2.101	126.2

TABLE 4.02

THE FREQUENCY DISTRIBUTION OF THE MASSES OF AGGREGATES OF STARS

Relative total mass range	log (mean mass) + constant	Number of aggregates	log (no./unit mass range) + constant
1-2	0.0	4	2.7
2-4	0.3	23	3.2
4-8	0.6	31	3.0
8-16	0.9	18	2.4
16-32	1.2	4	1.5
32-64	1.5	2	0.9
64-128	1.8	1	0.3
128-256	2.1	1	0

Table 4.01 gives the relative total masses of the clusters and associations, and the frequency distribution of the masses is given in Table 4.02 and Fig. 4.01.

Clusters of very low mass, containing few stars, are difficult to find and their lifetimes are relatively short in the face of disrupting forces. Consequently the tabulated numbers of them are doubtless a gross underestimate, and the point at log M = 0 on Fig. 4.01 has little weight.

Clusters of large mass are rare, and the two or three points of highest mass on Fig. 4.01 are of little value on account of the small number of objects involved.

With these qualifications, the points are then fitted with a straight line of slope - 2.2 $\pm$ 0.12 (error estimated by inspection). Thus the frequency distribution of the masses of aggregates of stars may be represented over most of its range by

$$\psi_a(M) \propto M^{-2.2 \pm 0.12} \qquad (4.01)$$

where $\psi_a(M)$ is the number of aggregates with masses in unit range of mass centred on M.

The similarity of the index to the value - 2.35 applicable to the mass function for stars (chapter 3) indicates that the process by which whole aggregates of stars form may be similar to that which produces the individual stars.

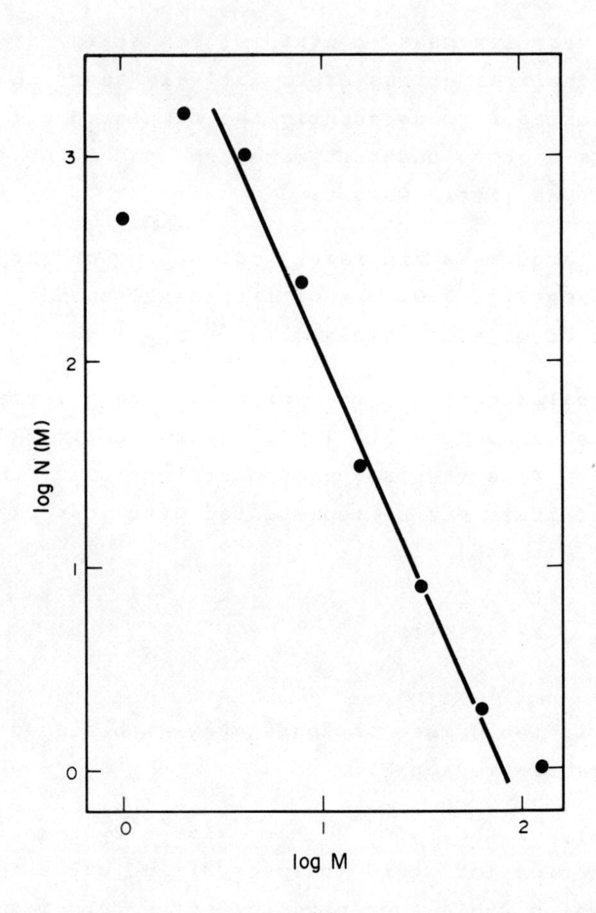

Fig. 4.01
Frequency distribution of masses of clusters and associations. N(M) is the relative number in unit range of mass at mass M.

Chapter 5.

THE FREQUENCY DISTRIBUTION OF CLOUD MASSES.

If stars form by the contraction of the interstellar clouds observed to be distributed in the Galaxy, the frequency distributions of the masses of the stars and of the clouds should be related. On the other hand, if stars form as an immediate consequence of the fragmentation of relatively massive clouds, we should compare the frequency distribution of the masses of aggregates of stars - clusters and associations - with that for interstellar clouds.

Data on the masses of clouds come from three kinds of observation - interstellar absorption lines, especially CaII, 21 cm hydrogen emission lines, and obscuration by dust.

5.01 Interstellar CaII Absorption Lines

The primary source of data is the catalogue by Adams (1949) of lines in some 300 OB stars. A detailed interpretation of the frequency distribution of the CaII absorption line depths, in terms of the frequency distribution of the masses of the clouds containing the absorbing gas, has been made by Penston *et al*, (1969). The interpretation depends on the assumption that the clouds are discrete and spherical: an aggregate of unresolved clouds would be treated as a single cloud.

Penston *et al* make allowances for the greater probability of finding larger clouds, and for the effect of the line of sight passing through the cloud at any random position. They finally arrive at the distributions of cloud masses shown in Fig. 5.01 where the zero point of the abscissa is determined by assuming that the number density of hydrogen atoms in the clouds is $n_H = 30 \text{ cm}^{-3}$ and $n(\text{CaII})/n_H = 1.2 \times 10^{-6}$. Over the range $20 \leqslant M/M_{\odot} \leqslant 2 \times 10^3$ the number of clouds per unit range in mass is given closely by

$$\psi(M_c) \propto M_c^{-3/2} \tag{5.01}$$

At higher masses the number of clouds lies below this relationship, apparently falling steeply towards zero for $M_c/M_\odot > 10^4$.

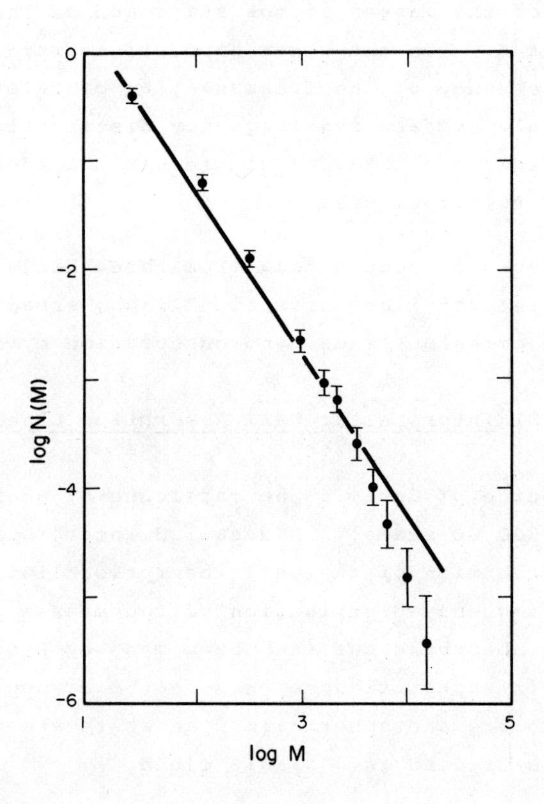

Fig. 5.01
Frequency distribution of masses of interstellar gas clouds. N(M) is the relative number in unit range of mass at mass M.

5.02 21 cm Hydrogen Line Emission

A detailed survey of the cloud structure of interstellar neutral hydrogen, in the region $100^{o} \leqslant l^{II} \leqslant 140^{o}$, $13^{o} \leqslant b^{II} \leqslant 17^{o}$, has been made by Heiles (1967a) using high angular and velocity resolution, 10 arc min and 1 km s^{-1} respectively. Earlier surveys (see Heiles, 1967a, for a bibliography) had been interpreted in terms of a cloud model, but the information was generally crude and gave at best average and uncertain cloud parameters.

Heiles found that three-quarters of the emission is due to gas in a smooth, diffuse background. It was estimated that most of the gas is at a distance of about 500 pc; the angular resolution therefore sets an upper limit of about one parsec on the scale of any structure occurring in this background.

Some 15% of the gas is in streams moving along the galactic spiral arm, and showing some structure, including aggregates of clouds. The density in these aggregates is only about twice that in their surroundings.

The remaining 10% of the gas is in true clouds of about stellar mass, having radii of a few parsecs and densities which are about twice as high again.

Heiles quotes Field as stating that "the mass spectrum of the cloudlets is well approximated by an inverse square power law over the range 6-20 $M_{\odot}$ consistent with Salpeter's (1955) initial luminosity function for stars which can also be fitted fairly well over the range 9-25 $M_{\odot}$ by an inverse square power law". That is,

$$\psi(M_c) \propto M_c^{-2} \tag{5.02}$$

A similar result is obtained by Perry and Helfer (1972) from an analysis of 21 cm H-line observations in the Perseus direction.

However, if stars form by the contraction of these clouds, an equality between the frequency distributions of the masses of the stars and of the clouds would be expected only if the cloud density is independent of cloud mass. This can be shown as follows (see also section 6.03):-

Let $\psi(M)$ be the number of stars formed per unit mass range about mass M, and let $\psi(M_c)$ be the number of clouds which exist in unit mass range about M_c.

The time scale for contraction of a cloud of density ρ is

$$t \propto (4\pi G\rho)^{-\frac{1}{2}} \tag{5.03}$$

and consequently the rate at which clouds contract to form stars is $\propto \rho^{\frac{1}{2}}$. Therefore

$$\psi(M) \propto \psi(M_c).\rho^{\frac{1}{2}} \tag{5.04}$$

Evidently a knowledge of $\rho(M_c)$ would be valuable in assessing the extent to which stars form by direct contraction of the observed distribution of clouds, and in understanding the rate and mechanism of cloud formation.

5.03 Dust Clouds

Heiles pointed out (1967a, 1969) that dark dust clouds show little if any HI emission, and he concluded that the hydrogen in these clouds is molecular. It seems to be most likely that stars form out of these dark clouds rather than out of the clouds discussed in the previous section; the observational evidence in chapter 2 shows a strong spatial relationship between dark clouds and young stars. Therefore the frequency distribution of the masses of clouds determined from observations of the distribution of interstellar dust may be the more relevant information. It is assumed in the interpretation of such observations that the ratio of gas mass to dust mass in the clouds is a constant. This assumption ought to be regarded with some caution in view of the indication that circumstellar dust clouds do not contain gas, although this may be the result of processes peculiar to circumstellar clouds only.

Various surveys of the sizes and distribution of interstellar dust clouds have been made, and the principle features of the more recent ones are summarised in Table 5.01. Peters' (1970) paper includes a bibliography of earlier work. Scheffler gives the number of clouds intersecting the line of sight, per kpc.

Let N denote the number of clouds intersected by the line of sight over a distance l, let a be the cloud radius and n the number of clouds per unit volume; then

$$n = N/\pi a^2 l \qquad (5.05)$$

This relationship has been used to derive the number of clouds pc^{-3} given in the fifth column of Table 5.01.

The opacity due to grains is

$$\pi r^2 Q_{ext} n_g d = \tau \qquad (5.06)$$

where r is the grain radius, Q_{ext} the extinction efficiency, n_g the number of grains per unit volume averaged over d, the depth along the line of sight, and τ is the optical depth.

Putting $\pi r^2 Q_{ext} = 10^{-10}$ cm^2, typical for interstellar grain models, then since $\tau \approx \Delta m$ we have

$$n_g \simeq 10^{10} \Delta m/d \qquad (5.07)$$

The mass of a single grain is $m_g = 10^{-15}$g, and so

$$\rho_g = n_g m_g \simeq 10^{-5}\ \Delta m/d \qquad (5.08)$$

Comparisons of obscuration to total interstellar mass density ρ, derived mainly from 21 cm measures of HI, lead to

$$\rho = 100\ \rho_g = 10^{-3}\ \Delta m/d \qquad (5.09)$$

and assuming this to hold for the clouds considered here, the mass of a cloud is then

$$M_c/M_{\odot} = 2.5d^2\ \Delta m \qquad (5.10)$$

where d is the cloud diameter in parsecs.

Masses derived from equation (5.10) are given in column 6 of Table 5.01. This small amount of data is fitted by the relationship

$$\psi(M_c) \propto M_c^{-2 \cdot 0} \qquad (5.11)$$

if it is assumed that the mass range is comparable with the average mass in each case.

The densities obtained from equation (5.09) are given in column 7 of Table 5.01. They can be related to the masses by the proportionality

$$\rho_c \propto M_c^{-0 \cdot 16} \qquad (5.12)$$

Equations (5.04), (5.11) and (5.12) then give for the frequency distribution of the masses which would form from these dark clouds

$$\psi(M) \propto M^{-2.16} \tag{5.13}$$

Having regard to the meagre amount of data and its uncertainty, the index is sufficiently close to that for stars (-2.35) and for aggregates of stars (-2.2) not to be inconsistent with the view that stars and whole stellar aggregates form by the condensation of the observed dark clouds, and that they are all formed by essentially the same process.

TABLE 5.01

Surveys of Interstellar Dust Clouds

Investigator	Method	Cloud				
		Diameter	Opacity	Number	Mass $M_{\odot}$	Density g cm^{-3}
Paczynski (1964)	star counts	7 pc	0.3 mag		37	$1.5.10^{-23}$
Neckel (1966)	interstellar reddening	100-1000				
Scheffler (1967)	interstellar reddening	3	0.26	5 kpc^{-1} $\Rightarrow$ 7 x 10^{-4} pc^{-3}	6	3.10^{-23}
		70	1.6	0.5 kpc^{-1} $\Rightarrow$ 1.3 x 10^{-7} pc^{-3}	20,000	8.10^{-24}
Schmidt (1968)	polarimetry	1-10		5 kpc^{-1}		
Fitzgerald (1969)	interstellar reddening	100-1000				
Peters (1970)	brightness fluctuations	10	0.37		90	$1.2.10^{-23}$

Chapter 6.

THE RATE OF STAR FORMATION.

Given the frequency distribution of the masses of stars forming - the initial mass function $\psi(M)$ - the rate of formation of all stars is determined if we know the rate of formation of stars in any mass range, or the rate of condensation of mass into stars. From the analytical point of view it does not matter which of these courses is followed, but from the point of view of the physics of star formation the alternatives may have very different implications, as follows.

If stars form simply by the contraction of interstellar clouds, then the frequency distribution of stellar masses will reflect the rate of formation of stars at each mass; it was pointed out in chapter 5 that this would be determined by the frequency distribution of cloud masses and by the time scale for contraction, which in turn depends on cloud density and which may itself be a function of cloud mass. If this is the process of star formation it is appropriate to investigate it by measuring the rate of formation of stars in a given range of stellar mass.

On the other hand, if stars form by the fragmentation of larger clouds into aggregates of protostars, then the frequency distribution of stellar masses is a consequence of the process of fragmentation, and the rate of formation of stars in any range of mass is governed by the rate of condensation of mass into stars of all masses through the fragmentation process.

In the first alternative the rate of star formation would be governed by the rate at which interstellar clouds of stellar mass formed and condensed; in the second case, it would be governed by the rate at which interstellar clouds reached the physical state necessary for fragmentation. A fragmentation process may occur in the first case, but not as a direct precursor of star formation.

However, given the initial mass function, chapter 3, either method of examining the rate of star formation can be related to the other. This will be done here; it does not presuppose any particular process by which stars form.

6.01 Some Analytical Relationships

Let dN be the number of stars formed in time dt in some volume or some system and let $\overline{M}$ be the average stellar mass. Then the rate of condensation of gas into stars is

$$-dM(gas)/dt = dM(stars)/dt = \overline{M}\ dN/dt \tag{6.01}$$

$\overline{M}$ at the present time is given by equation (3.07); if the stellar masses cover a large range as they are observed to do at present, i.e. if $M_2 >> M_1$, then that equation reduces to

$$\overline{M} \simeq 3.9\ M_1 \tag{6.02}$$

The rate of formation of stars in the mass range M-dM/2 to M+dM/2 is

$$\frac{dN}{dt}\ \psi(M)\, dM \tag{6.03}$$

where $\psi(M)$ is given now by equation (3.14).

The rate of formation of stars in the range of absolute magnitudes $M_v - dM_v/2$ to $M_v + dM_v/2$ is

$$\frac{dN}{dt}\ \psi(M_v)\, dM_v \tag{6.04}$$

where $\psi(M_v)$ is the initial luminosity function and is related to the initial mass function $\psi(M)$ by the initial mass-luminosity relation $M(M_v)$ as described in section 3.04.

It is to be expected that in general dN/dt and $\psi(M)$ (and consequently $\psi(M_v)$) will depend on the physical, chemical and dynamical conditions in the clouds from which stars are forming, and that these will change with time and position in any galaxy and will differ from galaxy to galaxy.

6.02 The Rate of Star Formation in Galaxies as a Function of Gas Content, from Bright Star Counts and 21 cm Hydrogen Line Measurements.

Beyond the local region of our Galaxy, and in all other galaxies, we can examine directly the rate of formation only of bright and hence massive stars. Any conclusions we draw regarding the rate of formation of all stars depends on a knowledge of the form of the initial mass function in these other galaxies and in other regions of our own Galaxy, which similarly cannot be determined directly. Estimates of the numbers of faint stars of low mass in galaxies are based, for example, on calculations of the total luminosity per unit mass of the galaxy as a whole, and on analyses of their spectra especially at red and near infrared wavelengths.

The total luminosity function of a galaxy at time t_1 is given by

$$N.\ \emptyset(M_v, t_1) = \int_0^{t_1} \frac{dN}{dt}.\ \psi(M_v, t, t_1)\,dt \qquad (6.05)$$

where N is the total number of stars in the system and $\emptyset$ and ψ are defined in chapter 3. $\psi(M_v, t, t_1)$ is the luminosity function at t_1 of stars all formed at t. It differs from $\psi(M_v, t)$ on account of the effects of stellar evolution during the time interval $t^1 = t_1 - t$; since this may have resulted in some stars having disappeared altogether (for example, as a result of supernova explosions), ψ must be normalised with respect to the number of stars formed at time t.

If equation (6.05) is restricted to stars with lifetimes $\tau < t_1$ then the lower limit of integration becomes $t_1 - \tau$. If dN/dt and ψ are changing slowly with time and we restrict ourselves to bright, massive stars for which $\tau << t_1$, equation (6.05) becomes

$$dN/dt = N.\ \phi(M_v,\ t_1) \Big/ \overline{\psi(M_v,\ t,\ t_1).\ \tau} \qquad (6.06)$$

If we also assume that the initial luminosity function is the same at the present time in each of the galaxies to be examined, and that stars in them with a given luminosity have a given lifetime so that $\tau(M_v)$ is a constant for any M_v, then

$$dN/dt \propto N \int_{(M_v)_b}^{\infty} \phi(M_v) \Rightarrow N_b \qquad (6.07)$$

and subject to the assumptions stated we can measure the relative rates of star formation in galaxies by inter-comparing counts of stars brighter than a given absolute magnitude $(M_v)_b$.

It is important that the absolute magnitude is chosen to exclude bright evolved stars of great age such as population II giants; this means choosing M_v brighter than -3.

Star counts have been made in a number of galaxies, and the data are listed in Table 6.01 together with measurements of the mass of neutral hydrogen in each system, obtained from the list compiled by Roberts (1969). The data in the Table are displayed in Fig. 6.01. They show a dependence of the numbers and hence of the rate of formation of bright stars on HI mass in galaxies given by

$$dN/dt \propto N_b(M_{pg} \leqslant -4.5) = 5 \times 10^{-6}\ (M(HI)/M_\odot) \qquad (6.08)$$

There is no indication of a significant dependence on total gas content; if any dependence exists Fig. 6.01 shows that it cannot affect the rate of star formation by more than about a factor two over a range in gas masses of 10^5 :1.

TABLE 6.01

System	Limiting M_{pg}	N_b	Reference	MHI $10^9\ M_\odot$	$\log \frac{N_b (M_{pg} \leqslant -4.5)}{(MH)}$
Sextans A	-4.0 (a)	343	Reddish, 1975b	0.070	-5.07
NGC 6822	-5.0 (b)	418	Kayser, 1966	0.15	-5.80
SMC	-4.5 (c)	3,000	"	0.48	-5.40
LMC	-4.7	4,700	de Vaucouleurs, 1955	0.54	-5.16
M33	-4.5	40,000	Reddish, 1975b	1.7	-4.63
M31	-4.5	120,000	Reddish, 1962c	8.5	-4.85
NGC 2403	-8.5	46	Tamman & Sandage, 1968	4.9	-6.03

Notes:

(a) assuming zero obscuration;
(b) limit is M_v = -5.0 but Fig. 5 in reference shows median C.I. ~ 0;
(c) assuming m - M = 18.7.

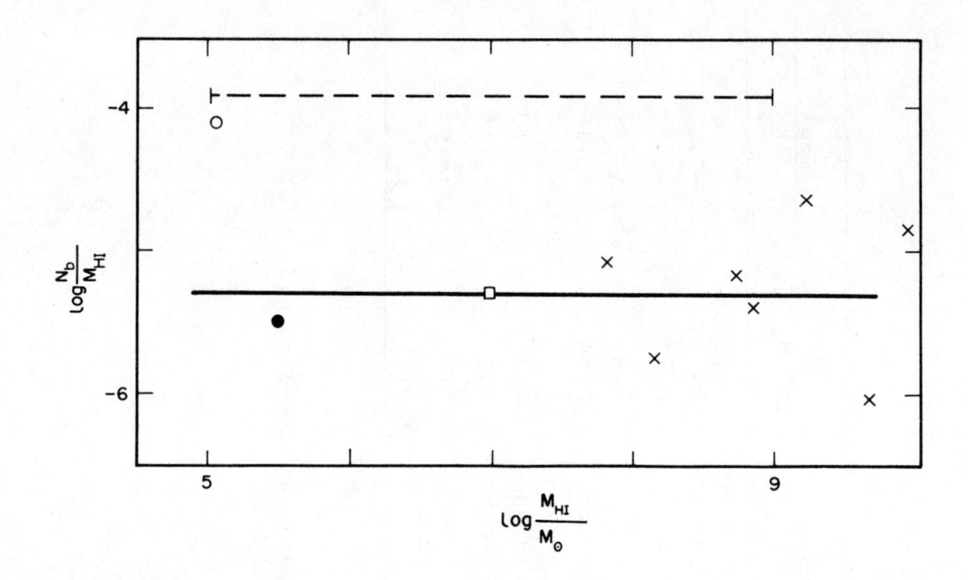

Fig. 6.01

Number N_b of stars brighter than $M_{pg} = -4.5$ against hydrogen gas content M_{HI}.

Crosses;	galaxies from Table 6.01
Filled circle;	solar region
Open circle;	Orion aggregate
Square;	Cygnus X region
Broken line;	M31, 0.5 to 1.5 arc deg from the centre
Continuous line;	mean value excluding Orion and M31

Further support for this view is provided by extensive counts of HII regions in galaxies (Hodge 1966), the numbers of which should be a measure of the rate of formation of bright stars; they show a one-to-one relation between the numbers and the HI masses.

The value of N_b for M31 has been corrected for obscuration within that galaxy (Reddish 1962c) whereas the values for other galaxies have not and may be too low by a factor of about two on that account.

Uncertainty is introduced also by lack of knowledge of the extent to which the HI mass is proportional to the total gas content, primarily on account of the possibility that a large fraction of the hydrogen may be in molecular form and that this fraction may itself be dependent on the total gas content of a galaxy.

Arguments are presented in chapters 7 and 8 which favour the view that a large fraction of the interstellar gas in our Galaxy is H_2, especially in the region 5 to 8 kpc from the galactic centre where star formation is most active. It is argued that the ratio H_2/HI depends steeply on total interstellar density and consequently that HI is most abundant in the less dense regions, and that its distribution within a galaxy is not typical of the interstellar gas as a whole. If these arguments are correct, then whether or not the mass of HI is representative of the total gas content of galaxies depends on whether or not the interstellar density in a galaxy depends on its gas content: it may more likely be expected to depend on the total mass of the galaxy.

These arguments also introduce additional doubt into any attempt to extend the star counts over a wider range of gas masses by using data relating to relatively small volumes of our Galaxy. Such a procedure is in any case uncertain because the rate of star formation per unit gas mass may be expected to vary from one region of a galaxy to another dependent on local physical conditions. However, the great increase in the range of MHI which can be achieved by considering local regions of our Galaxy makes a cautious examination of bright star counts and HI masses in such regions worthwhile.

(i) <u>Orion Aggregate</u> Menon (1958) has found the mass of HI to be 110,000 $M_\odot$ in a sphere of diameter 120 pc. Star counts by Sharpless (1952) give 46 stars brighter than M_v = -2 over an area of the aggregate 100 pc^2, and there are 82 stars of type B5 and earlier, that is, brighter than M_{pg} = -1. Using the luminosity functions given in Table 3.01 this corresponds to

9 stars brighter than $M_{pg} = -4.5$.

(ii) Cygnus X Davies (1957) gives an HI mass of 10^7 $M_\odot$ and 40 stars O9 and brighter, that is to say, brighter than about $M_{pg} = -5.1$, corresponding to 58 stars brighter than $M_{pg} = -4.5$.

(iii) Local Region of the Galaxy McCuskey (1956) gives the volume density distribution of B5 stars in the solar region; this averages about 1 per 10^5 pc^3 within 1 kpc. Taking an average thickness of 100 pc for the volume occupied by these stars, this gives 1 B5 star per 10^3 pc^2 in the local region of the Galactic disc. This may be taken as a measure of the number of stars in the magnitude range $-1.2 \leqslant M_{pg} \leqslant -0.8$ corresponding to 1 star in 8 x 10^4 pc^2 brighter than $M_{pg} = -4.5$.

The total amount of hydrogen per 10^3 pc^2 is 3.7 x 10^3 $M_\odot$ in this region according to the data given by Oort (1965a) but recent Lyα measurements of interstellar HI densities indicate that they may be lower by a factor 10 than those deduced from the 21 cm measurements.

The addition of these data to Fig. 6.01 confirms a relationship

$$N(M_{pg} \leqslant -4.5) = 4.8 \times 10^{-6} \, (MHI/M_\odot) \qquad (6.09)$$

over the wide range $10^5 < MHI/M_\odot < 10^{10}$, excepting the point for Orion; the high value in that case is to be expected for a region evidently unusually active in forming stars at the present time. Having regard to the comments made above regarding the need to correct some of the values of N_b for the effects of obscuration within the galaxies it is concluded that there is probably no significant dependence of N_b/MHI on MHI.

(iv) M31 Intercomparison of the surface density distributions of stars and gas along the south west major axis of M31 (Reddish, 1962b; Gottesman, Davies and Reddish, 1966) shows a close relationship between the two; over the region 0.5 arc deg to 1.5 arc deg from the centre of M31, the region where most OB stars occur, the data quoted give

$$N_b(M \leqslant -4.5) = (1.1 \pm 0.2) \times 10^{-4} \; (MHI/M_\odot) \qquad (6.10)$$

This value (Fig. 6.01) is close to that for Orion and higher by an order of magnitude than the value for the galaxy as a whole. This shows again the effect of restricting the data to a region particularly active in star formation; a large part of the gas content of M31 falls outside that region.

6.03 Dependence of the Rate of Star Formation on Gas Density

If stars form directly as a result of the fragmentation of interstellar clouds, then the rate of star formation would not be expected to show a smooth dependence on the interstellar gas density, but rather to show a discontinuous rise from zero to $(M(\text{cloud})/\overline{M}(\text{star}))/t(\text{cloud})$ at some critical density (where t(cloud) is the contraction time scale $\sim 1/\sqrt{(4\pi G\rho)}$). The rate of star formation in any given volume will therefore depend in the short term on the mass contained in clouds with densities higher than the critical density, and in the longer term (time scales >> t(cloud)) on the rate at which cloud densities increase from below to above the critical value. Therefore in order to make a meaningful comparison of the rate of star formation with gas density, it is necessary to know not only the mean density but also the distribution of cloud densities, and to understand the mechanism by which this distribution is set up.

Two further complications arise.

Firstly, if the rate of star formation does not depend linearly on gas density but varies as ρ^n with $n \neq 1$ then, since $\overline{\rho}^n \neq \overline{\rho^n}$ when $n \neq 1$ and ρ varies significantly about its mean value, it is necessary to know the distribution functions of interstellar densities, in both wavelength and amplitude, in order to determine n. Without such a knowledge, the discrepancy between the real and apparent values of n can be very large. This can be seen by a simple example, as follows.

Let ρ(cloud) denote the gas density within clouds of equal mass, ν(cloud) the number of clouds per unit volume, and dN/dt the rate of star formation per unit volume. Consider the dependencies of these parameters on height above the galactic plane, z.

$$\left.\begin{aligned} \text{Suppose}\quad \rho(\text{cloud}) &\propto z^{p} \\ \nu(\text{cloud}) &\propto z^{q} \\ dN/dt &\propto \big(\rho(\text{cloud})\big)^{m}; \end{aligned}\right\} \tag{6.11}$$

then

$$\left.\begin{aligned} \frac{dN}{dt}(z) &\propto z^{mp+q} \\ \bar{\rho}(z) &\propto z^{p+q} \end{aligned}\right\} \tag{6.12}$$

and

$$\frac{dN}{dt}(z) \propto (\bar{\rho}(z))^{\frac{mp+q}{p+q}} = (\bar{\rho}(z))^{n} \tag{6.13}$$

If m = 1, then n = 1 and there is no ambiguity in the conclusion that the rate of star formation is proportional to gas density. If, however, $m \neq 1$, this is no longer the case, If, for instance, p = 0 so that ρ(cloud) = constant, then n = 1 whatever the value of m; or if m = 3 and p = q, then n = 2 whatever the value of p and q.

Secondly, the distribution of young stars gives information about the distribution of material out of which stars form. There is substantial evidence (see chapters 2 and 7) that stars form in regions where most of the hydrogen is H_2. Consequently a comparison of the relative distributions of young stars and HI probably indicates the relative distributions of H_2 and HI. It has been argued (Schmidt, 1959) that the rate of star formation is proportional to $n_{HI}^{\ 2}$, which may be interpreted to indicate that $n_{H_2} \propto n_{H_1}^{\ 2}$, a not unreasonable conclusion, rather than to indicate a direct physical dependence of the rate of star formation on gas density.

Without direct information about both the density of interstellar H_2 and the numbers of young stars in regions of star formation, it is not possible to determine directly the dependence of the rate of star formation on gas density. Available data concerning the gas densities in regions where star formation is active, and which give some indication of the density required for star formation, are considered in chapter 7. Comparisons of the gross surface and volume densities of stars and of gas have been given by Sanduleak, 1969; Hartwick, 1971; and Madore, van den Bergh and Rogstad, 1974; but for the reasons given above conclusions about the rate of star formation must be drawn with care.

Chapter 7.

THE ENVIRONMENT OF STAR FORMATION.

Having discussed the locations of star formation, the rate at which stars form, and their mass spectrum, it is pertinent now to consider the physical conditions in regions of star formation so that we may seek to determine how they influence these other factors through the processes by which stars form.

The observational data discussed in chapter 2 indicate that stars form in dense, dark clouds, and that these are ordered into spiral arms forming the flattest subsystem in the galaxy. In these clouds the obscuration is very high and recent observations summarised in the next section indicate that the hydrogen in them is probably mostly molecular.

The question therefore arises as to whether or not the rate of star formation (which was considered in chapter 6) is dependent on the abundance of H_2 and if it is, what is the galactic distribution of interstellar H_2?

7.01 Molecular Hydrogen in the Galaxy

Rocket borne spectrometers detected the Lyman resonance absorption bands of H_2 in the direction of ξ Per (Carruthers, 1970b) showing that some 40% of the hydrogen in that direction is molecular. Laboratory experiments by Lee (1972, 1975) demonstrated that hydrogen would combine to form molecules on grains with temperatures $\leqslant$ 20K leading us to expect interstellar H_2 to be generally abundant and this was shown to be so when Spitzer et al, (1973) found that $n(H_2)/n(H)$ ranges from 0.1 when A_V = 0.3 magnitudes to 0.6 when A_V = 1.0, whereas the abundance is $< 10^{-7}$ when $A_V < 0.15$.

Prior to these discoveries and laboratory predictions there was much indirect evidence for the existence of large amounts of interstellar molecular hydrogen based on the discovery that in the more dense dark interstellar clouds the density of HI does not increase in proportion to the grain density as measured by the amount of obscuration. The effect was first noted by van de Hulst et al, 1954; they found that for a dark cloud in Taurus the intensity of the 21 cm line radiation by HI from the centre of the cloud was only 30% as great as it would have been if it had increased towards the centre of the cloud by the same factor as the obscuration. They suggested that the "missing" hydrogen had combined to form H_2. A similar effect was reported by Bok (1956) for a dark cloud south of ρ Ophiuchi, the grain density inside the cloud exceeding that in its immediate surroundings by a factor ten on average and a factor one hundred in the dense core, without any corresponding increase in HI emission; Bok suggested that some 90% of the mass of the cloud may be H_2.

More detailed studies of the phenomenon have been made by Garzoli and Varsavsky (1966), by Varsavsky (1968) and by Meszaros (1968) who show, Fig. 7.01, that as observations penetrate towards the denser parts of the dark clouds in Taurus and Ophiuchus the abundance of atomic hydrogen relative to grains, n_H/n_g, decreases progressively as the grain density n_g increases. Heiles (1967a, 1967b) has shown that this effect is restricted to the dark clouds and does not apply in general to the much less dense intercloud regions.

Pronik (1963) and Gosachinskii (1966) report a decrease in n_H/n_g by a factor of about two in the Omega nebula NGC 6618, and attempts to detect atomic hydrogen in the southern Coalsack have not been successful, n_H/n_g apparently being deficient by a factor $\geqslant 8$ (Kerr and Garzoli, 1968). Some difficulty in estimating the amounts of the deficiencies arises from uncertainty regarding the excitation temperature of the HI producing the 21 cm lines in dark clouds. However, comparison of 21 cm emission with Ly α absorption indicates that most of the HI occurs in clouds with excitation temperatures of only

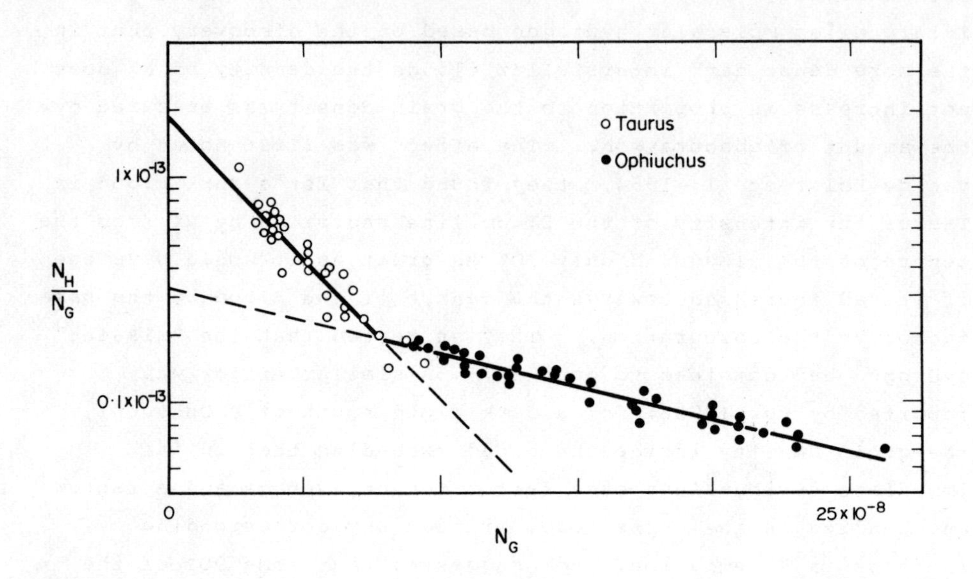

Fig. 7.01
Dust to gas ratio as a function of grain density. Filled circles, Ophiuchus; open circles, Taurus.

about 20°K as compared with the values 100°K formally adopted (Carruthers 1970a), so that the reduction in the excitation temperature in the coldest clouds is likely to be by less than a factor ~ 8.

X-ray observations of the Crab nebula indicate a column density of matter in the line of sight much higher than that deduced from 21 cm H-line measurements, suggesting large amounts of H_2 (Margon, 1974).

The existence of large amounts of H_2 is implied by the high abundance of other hydrogenic molecules, OH, NH_3, H_2O and CH_2O, in regions having relatively weak 21 cm HI emission (Heiles, 1967a; Feldman et al, 1969; Cheung et al, 1969; Synder et al, 1969; Sandquist, 1970). OH is generally distributed among dark clouds which show little or no HI emission (Heiles, 1970). The weakness or absence of HI emission in dark clouds may have been taken to suggest that there is little or no hydrogen associated with the dust were it not for the large abundances of hydrogenic molecules found in them. Measurements of 21 cm H-line radiation from 88 clouds give only 1% to 5% as much HI as would be expected from the OH or dust densities, indicating that 95% to 99% of the hydrogen in them has combined to form molecules (Knapp, 1974).

The strong infrared radiation from the centre of the Galaxy has been interpreted as radiation by dust grains (Lequeux, 1970: Okuda and Wickramasinghe, 1970). If this interpretation is correct, then according to Lequeux' calculation the ratio n_g/n_{HI} is ten times higher than the average value in the solar region, which may indicate that 90% of the hydrogen in the central region is molecular. Additional support for this conclusion comes from the observation that the ratio $n(\text{molecules})/n_{HI}$ increases progressively towards the Galactic centre (Robinson et al, 1964; Sandquist, 1970). The 2.6 mm radiation by CO in the central region of the Galaxy (Scoville and Solomon, 1974) indicates $10^7 - 10^8$ M⊙ of H_2 within 600 pc of the centre.

7.011 H_2 Densities

If the radius of the Taurus cloud is taken equal to the radius at which the obscuration has fallen to a half of its central value, then assuming that the distance to the Taurus cloud is 150 pc, the radius is 6 pc. Garzoli and Varsavsky (1966) give a column density 1 to 3 x 10^{21} H atoms cm^{-2} and since the rapid decrease in n_H/n_g with increasing n_g indicates that most of the hydrogen is in the cloud it follows that the average density of HI in the cloud is $n_H \sim 50\ cm^{-3}$. Fig. 7.01 indicates a reduction by a factor of from 2 to 10 in n_H/n_g as n_g increases from zero to its maximum value in the cloud, from which it is deduced that the density of molecular hydrogen reaches $n_{H_2} \sim 250\ cm^{-3}$.

In the centre of the Ophiuchus cloud, Bok concluded that the grain density is $\sim 10^3$ times the value outside the aggregate of clouds, and this would lead us to expect that $n_{HI} \sim 10^3\ cm^{-3}$, but most of the hydrogen appears to have combined, so that

$n_{H_2} \sim 500 \text{ cm}^{-3}$.

Heiles (1970) finds OH throughout dark clouds but, in the case of one dark cloud strongly illuminated by nearby OB stars, only in its centre. He derives the physical parameters

$$T_k \simeq T_e = 5.5^{\circ}K$$

$$n_{HI} \simeq 0.3 \text{ cm}^{-3}$$

$$n_{H2} \simeq 40 \text{ cm}^{-3} \text{ but possibly} > 10^3 \text{ cm}^{-3}$$

$$\text{mass motions} \sim 0.8 \text{ km s}^{-1}.$$

In the Sag B-2 cloud the thermal emission by interstellar ammonia indicates a density $n_{NH_3} \sim 3 \times 10^{-3} \text{ cm}^{-3}$, and the strength of the emission indicates collisional excitation by an ambient gas of density $n_{H_2} \geqslant 10^3 \text{ cm}^{-3}$, the 21 cm line emission indicating $n_{HI} < 3 \times 10^{-2} \text{ cm}^{-3}$.

Regions active in star formation, especially compact HII regions, emit radiation strongly in the maser-excited lines of molecules, and these regions generally have average densities $10^2 \leqslant n_e \leqslant 10^4 \text{ cm}^{-3}$, (Schraml and Mezger, 1969). The densities may be much higher locally in such regions. This does not necessarily mean that the molecules were formed before the stars, but it does imply that stars and molecules form in gas clouds with these densities.

Measurements of H_2CO and CO radiation from a molecular cloud near W51 (Scoville and Solomon, 1972) have been used to estimate densities of H_2 in the cloud. The average is $2 \times 10^2 \leqslant n_{H_2} \leqslant 10^3 \text{ cm}^{-3}$, probably rising to $\geqslant 10^5 \text{ cm}^{-3}$ near the compact HII regions of which there are three in the cloud. The cloud mass is $10^4 - 10^5$ M☉ of H_2 and a tenth as much HII.

In summary, atomic hydrogen appears to combine to form molecules in cold, dark clouds with densities $n_{HI} \geqslant 2 \times 10^2$ cm^{-3}, and stars apparently form from clouds of similar or higher density. The implication follows that star formation probably takes place in clouds of molecular hydrogen.

7.012 The Gross Galactic Distribution of Interstellar H_2

It was shown in chapter 6 that taking galaxies as a whole, the rate of star formation in a galaxy is closely proportional to the total mass of HI contained in it. On the other hand, in this section so far, evidence has been gathered which indicates that stars form from clouds of H_2. The evidence for the rate of star formation in galaxies comes from counts of OB stars and of HII regions, and that for HI masses from 21 cm line radiation measurements. The number counts and the HI masses cover a range of about five orders of magnitude. Consequently, providing that the ratio $M(H_2)/M(HI)$ in galaxies does not cover a comparably large range, it follows that the rate of star formation must be fairly closely proportional to the mass of H_2. In view of the evidence of star formation from H_2 it seems likely that the rate of formation is directly proportional to the mass of H_2 rather than to the mass of HI.

If we now consider in more detail the distribution of bright stars, HII regions and HI within galaxies, we may expect them to be similar to each other. This is indeed the case in M31, for instance (Gottesman et al, 1966; Gottesman and Davies, 1970); but it is not the case in some Sc galaxies, of which M33 is an example, and it appears not to be the case in our own Galaxy (see section 2.021). The dependence of HI and HII densities on distance from the centre of our Galaxy is shown in Fig. 7.02.

The peak in the HII density is associated with, and probably largely representative of, a peak in the distribution of compact HII regions (see section 2.013), the numbers of which are a measure of the rate of formation of OB stars. The close proportionality between the rate of star formation and HI mass in galaxies as a whole, described above, therefore leads us to expect that the amount of HI will vary with distance from the centre of the Galaxy in a manner similar to the HII density. However, as is seen on Fig. 7.02, the HI density distribution is very different from that of the HII distribution and peaks much further out. It is here suggested that most of the hydrogen in the region about 5 kpc from the centre is molecular and that this is the reason for the deficiency of HI.

Several observations discussed earlier in this section and in sections 2.01 and 2.02 support this suggestion. Dense concentrations of hydrogenic molecules such as OH, and dense dust clouds, are associated with compact HII regions.

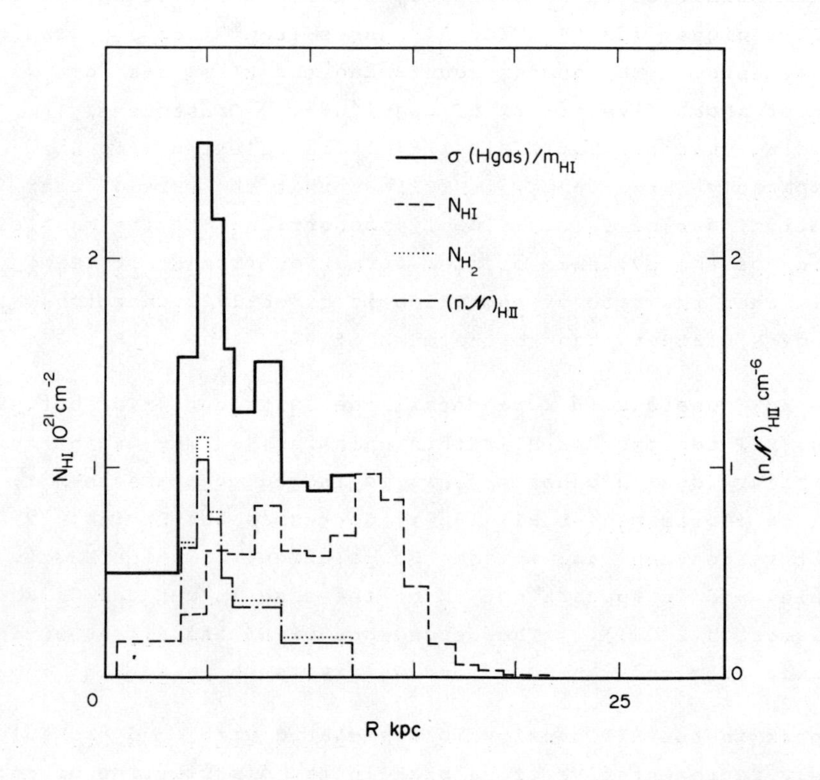

Fig. 7.02
Surface densities, at distances R from the centre of the Galaxy, of neutral atomic hydrogen, molecular hydrogen, ionised hydrogen and all hydrogen.

Compact HII regions are concentrated into the spiral arms from about 4 kpc to 6 kpc from the centre of the Galaxy, the arms appearing to have particularly large obscuration by dense clouds of interstellar grains. There is much evidence that the hydrogen in dense dark clouds is molecular. It also appears that the ratio n(molecules)/n_{HI} increases towards the centre of the Galaxy.

If we suppose that this suggestion is correct, we can estimate the amount of interstellar H_2 in the Galaxy.

The HII distribution shown in terms of (nN) cm^{-6} on Fig. 7.02 is similar to that of the giant HII regions (Burke, 1968), 50% of the emission, and of the regions, occurring between 4 kpc and 6 kpc from the Galactic centre. The HII distribution on the Figure will therefore be taken to represent the relative distribution of the rate of star formation as a function of distance from the centre of the Galaxy and hence, from the earlier arguments, of the relative amount of H_2. At 14 kpc from the centre the absence of HII indicates the absence of any substantial rate of star formation and hence of clouds of H_2; it follows that, with the addition of helium in a constant ratio to hydrogen, the HI density there measures the total gas density. If we now suppose that the total gas density is approximately constant in the region 8 to 14 kpc from the centre, and that the reduction in the HI density with decreasing distance from the centre results from the combination of some of the HI to form H_2 (the HII emission indicates a relatively low rate of star formation in that region), the distribution of H_2 can be drawn with the assumption that its amount is proportional to the rate of star formation indicated by the HII emission. The surface densities of HI, H_2 and HII, and of the total of hydrogen in all forms, is shown on Fig. 7.02. Integration over the disc of the Galaxy then gives

$$MHI = 3.6 \times 10^9 M_\odot = 2\% \ M(\text{Galaxy})$$

$$MH_2 = 2 \times 10^9 M_\odot = 1\% \ M(\text{Galaxy})$$

It may be concluded that it is consistent with the data on the Galactic distributions of HI, HII, compact HII regions and dark clouds, and with what is known of the detailed relationships between them, to suppose that about a third of the hydrogen in the Galaxy is in molecular form, that hydrogen molecules are largely confined to dark clouds with densities $n_{H_2} > 10^3$ cm^{-3}, that it is in such clouds that star formation occurs, and that the rate of star formation and the quantity of molecular hydrogen reach a maximum about 5 kpc from the Galactic centre, where up to ninety percent of the hydrogen has combined to form molecules.

7.02 Interstellar and Circumstellar Obscuration

7.021 The Obscuration in Dark Clouds

It has been noted that the densities in dark clouds are comparable with those in regions active in star formation, such as the compact HII regions. These latter also exhibit very high obscuration within them, and it is possible that dust grains play an important role in processes of star formation. For example, their effect on the rate of molecule formation may be crucial.

Interstellar obscuration reaches high values in the region 4 to 6 kpc from the Galactic centre, where star formation appears to be most active. Lynga (1968) and Westerlund (1969) both find obscuration in the 'Norma' spiral arm at $l^{II} \approx 330^o$ reaching several magnitudes at least. Westerlund notes that HII regions detected by their radio emission cannot be observed at optical wavelengths in this region. Indeed, of 176 HII regions detected at radio wavelengths, most of which occur in this annulus a few kpc from the Galactic centre, only 19 have been observed at optical wavelengths (Altenhoff, 1968). The total visual obscuration between the Sun and the Galactic centre is probably about 25 magnitudes (Becklin and Neugebauer, 1968); if this is divided between the two intervening spiral arms it indicates an optical thickness $A_V = 12$ for each arm. This will be reduced if some of the obscuration is within the central region of the Galaxy, but since it is an average through the arms in one direction only there may be local regions within the arms having even higher obscuration. Indeed it is not impossible that it mostly occurs in one cloud in the line of sight.

This level of obscuration is markedly higher than in those dark clouds in the anticentre direction where the hydrogen appears to have combined to form molecules. Although McCuskey (1956) estimates an average obscuration in this direction $A_v \approx 2.3$ through 3 kpc, there are other indications of much higher obscuration, at least in some parts of the region. It should be borne in mind that surveys to a given limiting magnitude tend to discriminate against regions of high obscuration since the stars there may be so faint as to be below the limit of detection.

The Galactic discontinuity at $l^{II} = 140^o$ has been variously interpreted as the end of the Perseus spiral arm (Becker, 1964) or as the effect of a nearby dark cloud (Reddish, 1967a; Bruck et al, 1968). The latter interpretation is supported by the sharp change in the density contours for stars at distances of only a few hundred parsecs, which appear to run almost radially at $l^{II} = 140^o$ in McCuskey's analysis (Fig. 7.03). Such an effect would result from the presence of a dark cloud containing large grains giving wavelength independent obscuration, and hence not taken account of by McCuskey's analysis based on the assumption that $Av/E_{b-v} = 3$. The star density contours would be "straightened" by the addition of a nearby dark cloud with a wavelength independent optical thickness of 2.4 magnitudes, making $A_v = 4.7$ altogether in this direction.

Other observations suggest even greater obscurations. Galaxy counts in Hubbles' zone of avoidance' in the anticentre direction have been made on the Palomar Observatory Sky Survey prints (Wilkenny, 1969). They show only one definite and three possible galaxies in 144 deg^2, and obscuration in the range $3 \leqslant A_{pg} \leqslant 9$ magnitudes, depending on the types of galaxy, would

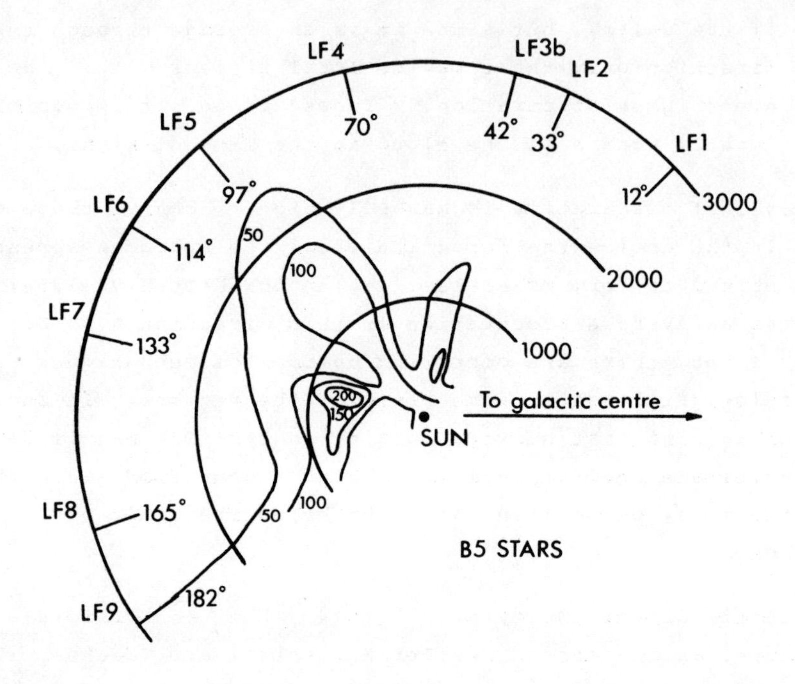

Fig. 7.03
The large-scale distribution in the galactic plane of the B5 stars. The numbers are in units of stars per 10^7 pc^3.

be required to reduce the expected numbers to this level. Inter-comparison of the relative distributions of OB stars and dark clouds (Sampson and Reddish, 1975) shows that no OB stars are visible through most dark clouds, indicating $A_v \geqslant 9^m$ in them. Assuming the depth of a cloud to be comparable with the thickness of the spiral arm, ~ 100 pc, and taking the optical cross-section of an interstellar grain to be 10^{-10} cm^2, it follows that $n_g = 3 \times 10^{-10}$ cm^{-3}, and $n_H \sim 300$ cm^{-3} if the usual ratio of hydrogen to dust is assumed (see earlier in this section). This is the density at which most hydrogen probably combines to form H_2.

In the Stock 2 area of Perseus a search for heavily reddened OB stars has revealed a number with colour excesses $E_{b-v} = 2.3$, indicating $A_v \approx 7^m$ (Pratt, 1975).

To summarise, the dark clouds along spiral arms have optical thicknesses typically in the range $7 \lesssim A_v \lesssim 12$ magnitudes, and there is some evidence that the obscuration is higher between the Sun and the Galactic centre than towards the Galactic anti-centre.

7.022 The Obscuration Surrounding Newly-Formed Stars

The recognition of compact HII regions, and the discovery that the exciting stars in them are generally so heavily obscured that they cannot be directly observed even in the near infrared, showed that newly formed OB stars are surrounded by dense clouds of dust grains (section 2.0133). It had previously been shown that many OB stars are embedded in circumstellar dust clouds (Reddish et al, 1966; Reddish, 1967c).

The failure to record the exciting stars in NGC 1491 ≡ CTB 12 on IN emulsions exposed to a limiting magnitude I = 16 (Reddish, 1968d) enables a lower limit to be set on the obscuration in the system. The Hα emission by compact HII regions indicates excitation by some tens of OB stars on average; in NGC 1491, the obscuration must be $A_I \gtrsim 14^m$, corresponding to $A_v \gtrsim 21^m$. Similar or greater amounts of obscuration must be present in most compact HII regions. A very heavily reddened cluster of OB and M supergiants in Ara (Westerlund, 1961) has been shown to have an obscuration $\gtrsim 13^m$ (Borgman et al, 1970).

The amount of obscuration of dust-embedded OB stars is up to $A_v = 12^m$ and dust embedded Wolf-Rayet stars have been found to have obscurations as high as $A_v = 10^m$ (Reddish, 1967c, 1968b).

The presence of circumstellar dust clouds around supernovae is indicated by the discovery (Seddon, 1969) that features in the spectrum of a supernova in NGC 2713 may be identified with interstellar absorption bands. The depth of the 4430Å band is $A(4430) = 0^m.5$ and since $A_{pg} \approx 20\ A(4430)$ this indicates a circumstellar obscuration of about ten magnitudes, similar to that surrounding the most heavily reddened OB and WR stars.

Whether these circumstellar clouds are the remnants of the clouds from which the embedded stars formed, or were produced by the stars during or after formation, is a matter for consideration in the second part of this book.

Compact HII regions and circumstellar clouds appear to have dimensions typically $\lesssim$ 1 pc, from which it follows that the number density of grains with optical cross sections 10^{-10} cm^2 necessary to give 10^m obscuration is $n_g \gtrsim 3 \times 10^{-8}$ cm^{-3}. If gas is present in the usual ratio to dust, $n_{HI}/n_g = 10^{12}$, then the gas density is equivalent to $n_{HI} \gtrsim 3 \times 10^4$ cm^{-3}.

7.03 Temperatures in Pre-stellar Clouds

The density in a cloud in which stars have formed may be expected to change with the speed at which a shock wave travels through it; that is to say, for clouds with dimensions 1 to 100 pc and shock wave speeds up to 100 km/s in HII regions, the density represents that at the time of star formation for times upwards of 10^4 year afterwards. In the case of temperature, however, the heating by transfer of radiation occurs so quickly that it is necessary for us to confine the investigation to clouds in which stars are about to form. Unfortunately we do not yet know how to distinguish such clouds and hence to select these for which measurements of temperature are relevant to our purpose. In these circumstances a summary of the known range of cloud temperatures is all that will be attempted.

The electron temperatures in HII regions are generally several thousand K, but recent measurements by radio interferometer have detected bright knots in the HII region W51 with brightness temperatures in excess of 10^5 K at a wavelength of 11 cm and with dimensions of 3000 a.u. or less (Miley, Turner and Balick, 1970). No spectra have yet been obtained so it is not known whether the emission is of thermal or of non-thermal origin. It may be that these nebulae are excited by newly formed stars within them, or they may themselves be stars in the process of formation. In the latter case, a point of particular relevance to the present study is whether or not the knots will suffer further sub-division into fragments of smaller mass; that is to say, whether or not they represent objects in which the process of star formation is essentially complete.

HI clouds generally appear to have temperatures of 60K to 80K (Hughes, Thompson and Colvin, 1971; Radhakrishnan and Goss, 1971).

Intercomparison of the strengths of interstellar line absorptions at 21 cm and at Lyα has led to the conclusion that much of the hydrogen is concentrated into clouds with a spin temperature of only 20 K (Carruthers, 1970a). Such clouds may be embedded in an intercloud medium (Clark, 1965) having density $n_H \sim 0.1\ cm^{-3}$ or less and a temperature $10^3\ K \leqslant T_K \leqslant 10^4\ K$.

At densities $n > 100\ cm^{-3}$, the kinetic temperature of the gas is given directly by the OH excitation temperature (Rogers and Barret, 1968; Goss, 1968). Dark clouds in which the hydrogen has combined into molecules have temperatures as low as 5 K (Heiles, 1970; Manchester and Gordon, 1971; Chaisson, 1974) and the possibility of even lower temperatures cannot be ruled out.

It is of interest also that 100 μm infrared radiation from the Galactic centre indicates dust grain temperatures $20 \leqslant T \leqslant 50K$ (Hoffman et al, 1971), and that Riegel and Crutcher (1972b) have found a cloud of HI some 100 pc from the Sun having $T \approx 20K$ and $n(HI) \approx 10\ cm^{-3}$ which they consider to be the region of the shock front of the local spiral arm density wave. In both cases the temperatures are at about the critical value (20K) for rapid formation of H_2 (Lee, 1972; 1975).

7.04 Densities in Pre-stellar Clouds

The known range of interstellar gas densities is summarised on Fig. 7.04.

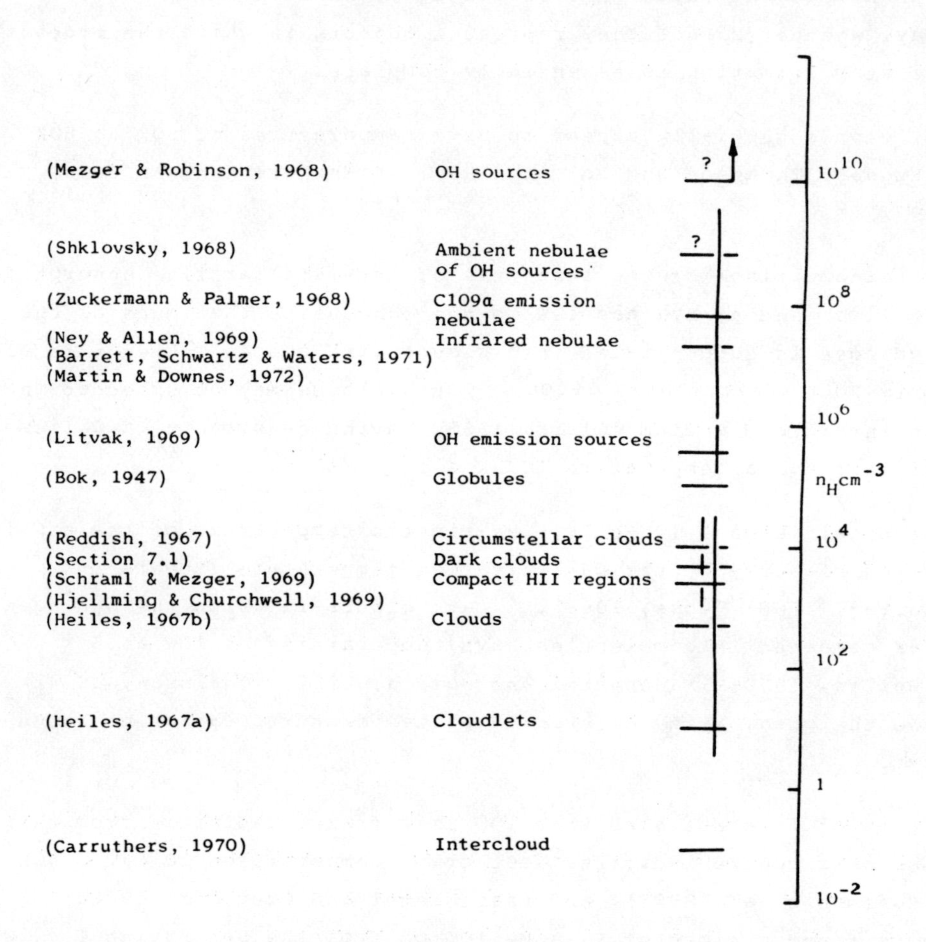

Fig. 7.04
Densities in the interstellar gas.

Compact HII regions appear to be evidence of the most recent star formation (Schraml and Mezger, 1969; section 2.013). They occur in greatest numbers in the regions of the Galaxy into which are concentrated the dense, dark interstellar clouds within which most of the hydrogen has apparently combined into molecules (section 7.01). It is probable that the gas densities in dark clouds and in compact HII regions are representative of values respectively before and after the epoch of star formation in a cloud. In both cases, average densities are typically $\geqslant 10^3$ molecules or atoms cm^{-3}, and may exceed 10^4 cm^{-3} locally. Of course it does not follow that this is the density at the time when a cloud fragments into protostar masses, and it is this density that we would particularly like to know. However, since the dark clouds are probably stable or contracting under the influence of external pressure and self gravitation, and the compact HII regions are expanding as a result of heating by their exciting stars, higher densities may presumably be reached during times in between these stages. It is therefore probable that the gas density exceeds 10^4 molecules cm^{-3} at the time a cloud fragments into protostars.

7.05 Interstellar Magnetic Fields

The strength of any interstellar magnetic field may affect the process of star formation in several ways; by confining charged particles, and through the pressure which it exerts, it may influence the cloud structure and stability; by linking a cloud with its surroundings it may change the motion or rotation of the cloud.

A description of the large-scale structure of the Galactic magnetic field within 1.5 kpc of the Sun has been given by Bingham and Shakeshaft (1967). It shows large-scale uniformity over this region, but with the occurrence of at least one major sharp reversal of direction.

From an analysis of measurements of the polarization of starlight, Schmidt (1968) concludes that there is an interstellar distribution of clouds having diameters of about 10 pc and containing magnetic fields of strength 10^{-5} gauss, distributed with a frequency of 5/kpc. Davies (1968b) analysed measurements of Faraday rotation of the galactic radio emission and concluded that there are two scales of cloud like fluctuations in the interstellar magnetic field structure, $\leqslant$ 3.0 pc and 0.5 to 0.05 pc.

Estimates of the strength of the general interstellar field distribution, from pulsar frequency dispersions (Smith, 1968) and from Faraday rotation (Bridle, 1969), are in the range 2 to 10×10^{-6} gauss. However, determinations by measurements of the Zeeman splitting of the 21 cm HI line indicate an upper limit of about 10^{-6} gauss in the nearby Orion spiral arm.

Much larger magnetic field strengths in dense clouds have been found by Zeeman measurements (Davies et al, 1968; Verschuur, 1969; Brooks et al, 1971), ranging from 10 to 50 micro gauss. These occur in HI clouds surrounding HII regions and it is suggested that they may result from field amplification due to compression or contraction of clouds in regions of star formation. A field strength of about 3 micro gauss has been found in the Taurus dark clouds.

A field strength reaching the extraordinary value of 6 gauss has been found in the molecular cloud in Orion A (Clark and Johnson, 1974). Given field conservation during compression according to the relationship $H \propto \rho^{2/3}$, starting with 10^{-6} gauss at $n \approx 0.1$ cm^{-3} would give $n \approx 10^{8}$ cm^{-3} at H = 6 gauss. Barret, Schwarz and Waters (1971) obtain a density $n_H \sim 2 \times 10^{6}$ cm^{-3} in Orion A, but there could also be large numbers of molecules or the density could be higher in the region for which the magnetic field was measured by the Zeeman effect in SO.

Viewed from the North Galactic pole, the magnetic field direction is predominately clockwise along the spiral arms (Verschuur, 1969).

7.06 Motions of the Interstellar Gas

The kinetic motions of interstellar gas clouds contribute to their compression by their forces of collision, and to their expansion through the energy they impart on collision and which becomes converted to heat. Such collisions also affect the rotations of the clouds and thereby the centrifugal forces acting on them. In these ways the motions of clouds influence their physical and chemical properties and hence indirectly the processes by which stars form. Furthermore, the motions of newly formed stars may be expected to reflect at least in part the motions of the clouds out of which they formed.

7.061 Large-scale Systematic Gas Motions

The rotation of a galaxy as a whole provides the dominant systematic motion of the interstellar gas. There is evidence, however, in our Galaxy and in M31 and M33, that the gas does not orbit with exactly the same speed as the stars most newly formed from it.

Dixon (1967) estimates that, at 10 kpc from the Galactic centre, the gas orbits with a speed 14 km s^{-1} less than gravitational circular velocity and is also moving radially outwards at a speed of 6 km s^{-1}. Stars are formed from the gas at apogalacticum and fall inwards, following orbits having eccentricities $\approx$ 0.1. The stellar velocity at apogalacticum is $V_a = V_o \sqrt{1 - e}$ where V_o is the circular velocity at the distance from the centre, about 250 km s^{-1}; consequently recently formed stars lag the circular velocity by about 12.5 km s^{-1} and the gas lags recently formed stars by 1.5 km s^{-1}. Oort (1965b) obtained a corresponding figure of 2 km s^{-1}, and Penston et al, 1969 obtain 3 km s^{-1}. From a comparison of the radial velocities of HII regions and their associated OB stars Murdin (1969) concludes that the gas and O stars are moving radially outwards relative to the B stars at an average speed of 7 km s^{-1}. These differential motions have been taken to indicate the presence of forces additional to gravitation - probably magnetic pressure - acting on the gas but not on the stars.

In M33, the gas in the spiral arms lags 15 km s^{-1} behind the disc of gas in which the arms are embedded, and has a higher velocity dispersion, $\pm$ 9 km s^{-1} as against $\pm$ 5 km s^{-1} (Carranza et al, 1968). There is also a marked difference, of 40 km s^{-1}, between the rotational velocities of the northern and southern arms.

The systematic motions of the HI gas and HII regions in M31 differ by less than 10 km s^{-1} if at all (Gottesman and Davies, 1970), although there is a difference of some 30 km s^{-1} between the rotational velocities of the two arms, within about 40 arc minutes of the centre. There may be a small radially outward motion, about 5 km s^{-1}.

In the "discrete cloud model" of the interstellar gas in our Galaxy, according to which clouds with diameters 15 pc and gas densities 8 atoms cm^{-3} are distributed at average separations of 40 pc in a much less dense medium, the r.m.s. line of sight velocity of the clouds is $\pm$ 9 km s^{-1} (see Allen, 1964 for a summary of data to 1962). Penston et al (loc. cit.), find that the velocity dispersion decreases with increasing cloud mass from $\pm$ 23 km s^{-1} for light clouds to 9 km s^{-1} for massive ones. However, it may be that a considerable part of these measured dispersions are due to local systematic motions rather than a true dispersion. The velocity dispersion of HI clouds in the solar region is $\pm$ 6 km s^{-1} (Falgarone and Lequex, 1973), and the random motions of 80 dark clouds of molecules have been found to average only 5 km s^{-1} (Minn and Greenberg, 1973).

Kerr (1969) notes that the most recent 21 cm hydrogen emission line surveys show deviations from circular motion ~ 10 km s^{-1} occurring throughout the Galaxy. There is a difference of this amount between the apparent rotation curves on opposite sides of the Galaxy in the regions 5 to 8 kpc from the centre. In the anticentre and Perseus regions major parts of the hydrogen are moving with radial velocities of - 20 to - 30 km s^{-1} (Linblad, 1967; Rickard, 1968). The true velocity dispersion of clouds may therefore be much less than 9 km s^{-1}; note that the disc component in M33 referred to above had a velocity of only 5 km s^{-1} and it may be that the higher value of 9 km s^{-1} found for the spiral arms in that galaxy resulted from the inclusion of large-scale radial motions of parts of the arms. Within our Galaxy, a velocity dispersion of only 2 km s^{-1} has been found for HI clouds embedded in more extended HI distributions having a velocity dispersion of 10 km s^{-1} (Grahl et al, 1968). In the solar neighbourhood, Venugopal and Shuter (1969) find a mean velocity dispersion in the HI of 6.5 km s^{-1}.

High velocities have been found among gas clouds at considerable distances from the galactic plane, but since these are not in regions where star formation appears to take place they will not be discussed here. A recent summary of available data is given by Dieter (1969). One aspect of them may, however, be particularly relevant to the present text. It is suggested (Oort, 1970) that they represent an inflow of intergalactic hydrogen at a rate of 10^{-11} of the mass of the Galaxy per year, and this is comparable with the rate at which interstellar gas in the Galaxy is condensing to form stars.

7.062 Motions within Clouds

7.0621 HII regions Most of the velocity data comes from two kinds of measurements - of high level recombination emission lines of hydrogen at radio wavelengths, and of the Hα emission line by pressure-scanned Fabry-Perot interferometers. The former have high velocity resolution and low angular resolution and the latter vice-versa; consequently there has been some difficulty in distinguishing between internal turbulence and systematic internal motions such as expansion, since the two are averaged over the beamwidth in the radio telescope observations and, although interferometers give better velocity resolutions than did earlier methods, $\approx$ 2 km s^{-1}, it is barely adequate and furthermore the Hα observations do not penetrate clouds of dust which obscure so much of the HII regions.

Radio observations generally show r.m.s. velocity dispersions of $\pm$ 25 km s^{-1} (Dieter, 1967; Mezger and Hoglund, 1967), and Hα observations of the brightest HII regions in the nearby external galaxies M33 and M101 show similar velocity dispersions, ~ 27 km s^{-1} (Smith and Weedman, 1970). On the other hand, very low velocity dispersions have been found in Galactic HII regions. Courtes et al (1968) find dispersions in about a hundred regions which average $\pm$ 6 km s^{-1}. Smith and Weedman (loc. cit.) obtain values ranging from $\pm$ 2 km s^{-1} to $\pm$ 5 km s^{-1} among five regions, including $\pm$ 2.8 km s^{-1} for M20 contrarily to Courtes et al. who give $\pm$ 11 km s^{-1} for this nebula. The reason for this disagreement may be found in the different parts of the nebula measured. Smith and Weedman (1971) find that the most probable internal velocity in 70 HII regions in the LMC is 15 km s^{-1}. Foukal (1969) finds a dispersion of $\pm$ 5 km s^{-1} in NGC 7000, in reasonable agreement with Courtes' $\pm$ 5.9 km s^{-1} and Smith and Weedman's $\pm$ 2.1 km s^{-1}; and he finds 6 km s^{-1} to 12 km s^{-1} in M20 in agreement with Courtes. In M8, however, for which the others find 3 km s^{-1} to 8 km s^{-1}, Foukal finds 10 km s^{-1} to 20 km s^{-1} and concludes that the HII region consists of a rapidly expanding core surrounded by a relatively quiescent outer shell. A similar result has been obtained by Meaburn (1971). Observations which do not include the rapidly expanding core of an HII region will generally give low velocity dispersions. Since the cores of the compact and most luminous HII regions are often heavily obscured, Hα measurements of them will give much lower velocity dispersions than the radio wave measurements which penetrate the dust. Foukal comments that smaller velocity variations are found where there is evident obscuration which limits observation to the outer parts of the HII regions.

An interesting example of an HII region with an expanding core is provided by NGC 2068 (M78). Spectra of the starlight scattered by the dust grains in the nebula indicate that the dust system is expanding with a velocity of about 100 km s^{-1}, but no similar expansion is shown by the Hα emission line from the gas (Stockton and Chapman, 1970). Dust appears to have been blown away from one region of the nebula altogether.

The expansion of the cores of HII regions in general, and this last example in particular, present a problem when considered in relation to the circumstellar dust clouds in which so many young OB stars are embedded (Reddish, 1967a); why have these circumstellar dust clouds not been blown away?

7.0622 Dark clouds Improvements in the angular and frequency resolution of radio wave measurements applied to HI and molecular emission and absorption spectra have resulted in very low internal velocities of motion being found in some dark clouds.

HI absorption spectra have good angular resolution on account of the small extent of the continuum source supplying the radiation; spectra generally show several components due to the presence of several clouds or parts of clouds in the line of sight between observer and source; Hobbs (1974) concludes that the sum of the velocities due to turbulence, rotation and sub-cloud motions is 1.8 km s^{-1}. Shuter and Verschuur (1964) find an average spacing of the velocities of components of only 1.6 km s^{-1}, suggesting that cold HI clouds possess internal turbulent velocities of only this amount or less. Clark (1965), using an interferometer, finds a similar value, about 1.3 km s^{-1}. Clark estimates the sizes and total masses of his clouds and from them calculates the escape velocities; these average 0.6 km s^{-1} so that the internal motions must be less than this if the clouds are stable. Internal motions of about 0.5 km s^{-1} would be required for the stability of Bok's globule 'Barnard 227' which is estimated to have a mass of 30 $M_{\odot}$ and a diameter of 0.8 pc (Bok et al, 1970).

Internal speeds of motion probably as low as these escape velocities have been found in clouds from which OH line radiation has been detected; the lines usually resolve into several components which have widths and separations corresponding to about 0.75 km s^{-1} (Mezger, 1970). It is of relevant interest that dispersions < 2 km s^{-1} have been found for T Tauri stars in Taurus-Auriga (Herbig, 1962), and the velocity dispersion of stars in the Pleiades is only 0.7 km s^{-1} (Jones, 1970).

More detailed maps of the velocity fields inside dark clouds will probably become available from measurements of the 6 cm emission by molecules of formaldehyde. The short wavelength leads to narrow radio telescope beamwidths giving three times better angular resolution than OH radiation observations, and the molecule is widespread among interstellar clouds.

Interferometric measurements of the water vapour 1.35 cm emission have shown that it originates in sources of very small size, probably less than 1 a.u.,(Burke et al, 1970) and spectrometry shows a pattern of radial velocities which has been interpreted as indicating orbital motions of protostars (Knowles et al, 1969). A similar interpretation of the groupings of radial velocities of OH emission sources has been made by Shklovsky (1967, 1968).

7.07 Element Abundances in the Interstellar Gas

The condensation of interstellar gas to form stars, the processing of the gas by nuclear synthesis in stars and the subsequent ejection of processed material, for example by supernovae, link star formation to the composition of the interstellar gas in an intimate manner. Stars may be expected to possess initially the chemical composition of the interstellar clouds from which they formed, and by determining the compositions of stars of all ages and locations in the Galaxy we may hope to build up a picture of the history of the composition of the interstellar gas. The available data on stellar compositions is considered in section 7.08 but it is important also to determine the composition of the interstellar gas directly; partly because it adds to the general fund of information about the abundances of the various elements; partly because it is important to check the assumption that stars form with the same composition as the interstellar clouds - that during the process of star formation there is no separation of the elements by diffusion and no selective concentration by accretion processes. Information about the element abundances in the interstellar gas comes mostly from analysis of emission line spectra of HII regions with the advantage that these can be observed to greater distances in our own and other galaxies than can stellar absorption line spectra.

7.071 Helium in the Galaxy

Probably the most extensively observed gas cloud is the Orion nebula; recent analyses of optical emission line spectra agree on a value N(He)/N(H) = 0.11 $\pm$ 0.01 (Robbins, 1970; Peimbert and Costero, 1969). The latter authors find similar values for the nebulae M8 and M17 in the Sagittarius spiral arm, indicating no substantial change in helium abundance with a change in distance from the centre of the Galaxy by about 2 - 3 kpc.

Lower values than these are found by radio wavelength measurements of the H_e 109α/H109α line strength ratio. Mezger et al, (1970) find the average value in seventeen HII regions to be N(He)/N(H) = 0.090; however, the contribution of HeI is set at 0.002, whereas Peimbert and Costero obtain the larger value 0.01.

It may be concluded that the present abundance of helium in the interstellar gas in the local region of the Galaxy is in the range N(He)/N(H) = 0.09 to 0.12. The He abundance in planetary nebulae is higher, averaging 0.13 $\pm$ 0.02 (Harman and Seaton, 1966); since planetary nebulae eject in total some 10^{-2} $M_{\odot}$ yr^{-1} into the interstellar medium they evidently contribute, to a small extent, to the enrichment of the interstellar gas with helium.

7.072 Helium in other Galaxies

The brightest HII region visible in an external galaxy is 30 Doradus in the LMC. Optical observations give N(He)/N(H) = 0.08 (Faulkner and Aller, 1965; Wares and Aller, 1968), but radio observations (Mezger et al, 1970) give the much higher value 0.17 $\pm$ 0.06. It may be that the radio observations give higher weight to a dense central core by reason of their ability to penetrate deeper into the cloud; if so, this is considerable evidence in favour of enrichment of the interstellar gas by helium synthesised in massive stars such as excite the 30 Doradus nebula. The high helium abundance in the crab nebula, N(He)/N(H) = 0.45 (Williams, 1967) also shows that substantial amounts of helium are ejected by supernovae and it may be that there has been such ejection inside 30 Doradus.

Optical observations of emission regions in several other galaxies indicate helium abundances essentially the same as in our Galaxy (Welch, 1970; Peimbert and Spinrad, 1970a, 1970b) averaging 0.10, although the latter find the higher value 0.13 in NGC 604 in the Sc spiral galaxy M33 and a lower value 0.08 in NGC 346 in the SMC - the same as the optical determination for 30 Doradus.

Observations of HII regions in the Sb spiral galaxy M31 show a helium line strength decreasing by a factor 4 as the distance from the centre increases from 3 - 20 kpc (Ford and Rubin, 1968); it now seems most likely that this represents at least in part a real abundance effect and not just a change in excitation conditions (Peimbert, 1968). Observations of the spiral galaxy M101, however, show no dependence of helium abundance on distance from the centre (Peimbert and Spinrad, 1970b).

The nuclei of Seyfert galaxies show a high abundance of helium, $N(He)/N(H) \simeq 0.14 \pm 0.05$; on the other hand, it has been suggested that quasars are relatively deficient in helium by an order of magnitude (Osterbrock, 1970).

7.073 Other Elements

The abundances of nitrogen and oxygen in local HII regions in the Galaxy (Rubin, 1968; Peimbert and Costero, 1969) do not differ significantly from the solar values $N(O)/N(H) = 5.9 \times 10^{-4}$ and $N(N)/N(H) = 8.5 \times 10^{-5}$ (Lambert, 1968). Observations of other galaxies, however, indicate substantial variations. The nitrogen abundance appears to increase towards the nucleus of M31 (Ford and Rubin, 1968). Nitrogen is over-abundant by a factor 2 to 6 in the nuclei of the spiral galaxies M51 and M81

(Peimbert, 1968) and the irregular NGC 5253 (Welch, 1970) and in the HII regions in M101 it is deficient by a factor 4 in the outer parts and over-abundant by a factor 3 in the inner, in comparison with the 'standard' composition (Searle, 1971). Seyfert galaxies show still larger over-abundancies, that in NGC 3031 being 6 to 10 times the solar value (Alloin, 1970).

On the other hand, the irregular galaxy NGC 6822 is deficient in nitrogen by a factor 6 in comparison with the Orion and M17 nebulae in our Galaxy, although the helium abundance does not show any difference (Peimbert and Spinrad, 1970a). The variation in the nitrogen abundance with distance from the centre of a galaxy appears to be similar to that of metal abundance in the stars.

The situation is less clear in the case of oxygen. The abundance in the HII region NGC 604 in the spiral galaxy M33 is similar to that in Orion (Peimbert, 1970), whereas in the irregular galaxy NGC 6822 it is lower by a factor 1.7. In M31 there are large variations in the O/Hα spectrum line strength ratio at all distances from 3 - 20 kpc from the centre (Ford and Rubin, 1968) whereas in M101 there appears to be a slight inward increase in O/H (Peimbert, loc. cit.). The apparent variations in abundances deduced from variations in the line strength ratios must be regarded with some caution. The line strength ratio Hα/NII in Orion decreases substantially from the centre outwards due to variations in T_e and n_e (Baudel, 1970) and some of the effects observed in other galaxies may be due in part to variations in excitation conditions; but it is fairly certain that real abundance differences also occur in amounts at least comparable with those indicated (Peimbert, 1968).

Another factor, to be borne in mind when intercomparing abundances derived at times some years apart, is that calculated collision excitation strengths of positive ions of N, O and N_e have increased by some 50% in the last decade, giving a corresponding decrease in calculated abundances (Saraph et al, 1969).

Calcium, titanium and beryllium are deficient in the interstellar gas, relative to solar abundances, by factors 10 to 10^3 (Routly and Spitzer, 1952; Herbig, 1968; Habing, 1969). It has been suggested (Wickramasinghe, 1968) that these elements may be ejected from stars into the interstellar medium in solid form. It is possible that substantial fractions of the interstellar gas are captured by the grains in cold clouds - certainly the high abundances of molecules there indicates that this must be so if the molecules are indeed mostly formed on grain surfaces - but in the hot conditions of HII regions, for example, in Orion, M3 and M17, the gaseous abundances of N, O, Ne and S are similar to solar abundances although the dust contents appear to differ greatly among these nebulae.

7.08 Element Abundances in the Stars

The dependence of element abundances on the ages and locations of stars gives information about the composition of the gas, out of which the stars formed, in different parts of the galaxy throughout its evolution. If the processes of star formation depend on the composition of the material from which they form - and it would be surprising if they did not - this information is of direct importance in the present context. It has indirect value also, in that it may indicate that different types of stars must have been formed at different times or in different locations, thereby requiring an understanding of the reasons for these differences. For example, it appears that the present helium abundance can only be accounted for by either synthesis in one large initial cosmic event, or in the explosion of supermassive objects in the early stages of evolution of galaxies.

7.081 Stellar Helium Abundances

7.0811 The Sun During the last three years, determinations of the helium abundance in the Sun have given lower values than those previously accepted. An abundance N(He)/N(H) = 0.062 ± 0.007 now appears to be fairly well established (Lambert,

1967; Peimbert and Costero, 1969; Mitler, 1970). Since the present helium abundance in the interstellar gas in the solar region of the Galaxy (section 7.07) is 0.105 $\pm$ 0.015 it follows, if the solar composition is representative of the material from which the Sun formed, that the helium abundance has increased by a factor 1.7 since the Sun was formed. Taking the ages of the Sun and Galaxy to be 4.5 x and 10 x 10^9 yr respectively, this is a time increase by a factor 1.8.

7.0812 Other stars It is instructive to begin by considering a possible implication of the figures given in the preceding section, since they are probably the most reliable measurements of helium abundances and ages available to us. This implication is that the interstellar helium abundance in the Galaxy has increased linearly with time from an initial value not significantly different from zero; Fig. 7.05.

Available data covering the helium abundances in stars with various ages and locations may then be examined within the framework of the question "is there evidence that the initial helium abundances in stars depart significantly from the relationship shown on Fig. 7.05?"

Analysis of the spectra of 94 Population I (young) stars has led to an estimate of the helium abundance in them $N(H_e)/N(H) = 0.115$. However, the uncertainty inherent in deriving helium abundances from spectral line strengths in early type stars, as a consequence of departures from local thermodynamic equilibrium, has been emphasised by Hearn (1970), who concludes that the derivations are meaningless.

An even more uncertain situation is presented by spectral analysis of Population II (old) B stars, since they have reached that spectral type through a lengthy evolutionary process in which the element abundances in the interior of the star have certainly been changed drastically through nuclear synthesis, and the possibility of mixing between interior and envelope cannot be ruled out at all stages of evolution. Therefore

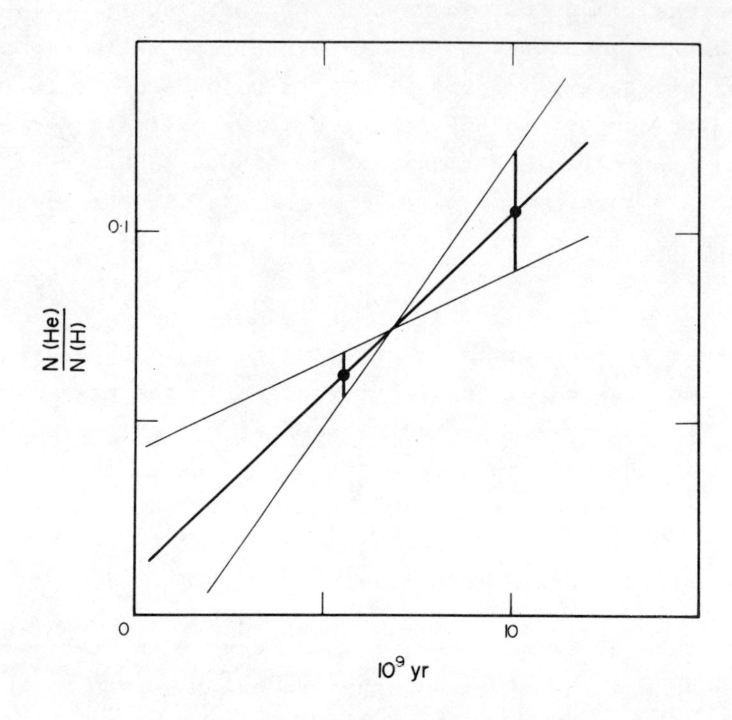

Fig. 7.05
Observational data on the growth of helium abundance with time, in the Galaxy.

apparently high atmospheric helium abundances in such stars do not necessarily represent those in the gas out of which the stars formed. A summary of the conflicting data has been given by Pagel (1970a). On the other hand, apparently very low helium abundances in old B stars (loc. cit.) are also of doubtful validity. For comparison i Ori B, a B star companion of an O star in the Ic Ori association with an estimated age of only 3×10^6 yr and therefore expected to have a helium abundance $\approx$ 0.1, has an abundance apparently only a tenth of the expected value, a result which may be due to the non-LTE affects referred to above (Conti and Loonen, 1970).

Another method for determining helium abundance is to make use of the dependence on chemical composition of the luminosity per unit mass of a star. The helium abundance of seven young pairs of stars derived in this way (Popper et al, 1970) is $N(H_e)/N(H) = 0.12^{+0.02}_{-0.03}$. A similarly derived abundance for the very old star pair μ Cas (Hegyi and Curott, 1970) gives $N(H_e)/N(H) = 0^{+0.05}_{-0.00}$.

These values are consistent with the relationship on Fig. 7.05, but the uncertainties in them are such that they cannot be taken to confirm it.

7.082 Other Elements

Interest has been centred mainly on the CNO group of elements, partly on account of their role in the CNO cycle of energy production synthesising helium from hydrogen in stellar interiors and partly because their spectrum lines are prominent; and on the iron group of elements, the term "metal abundance" having been used loosely to represent sometimes the abundance of iron and sometimes the average abundance of several heavy elements.

The available data will be considered in three respects, dependence on location, on time, and relative variations of abundances among the elements.

The metal abundances in stars in the Hyades cluster are greater than in the Coma cluster and in the Pleiades by a factor x 1.55 ± 0.15, although there are no significant differences among the stars within each cluster (Nissen, 1970b; Chaffee et al, 1971; Barry and Cromwell, 1974). Since the clusters have approximately the same ages, this indicates differences in metal abundance by this factor over distances of only 100 pc at a given time. Again, Hesser and Henry (1971), show that the calcium abundance in the galactic cluster M7 is twice the average among field stars but the variation within the cluster is only ± 25% indicating that it formed from a cloud of relatively homogeneous composition.

Estimates of metal abundances through the measurement of spectral line blanketing, using broad band photometry, has been a powerful tool for the investigations of abundance variations.

Analyses by Dixon (1966) have led him to conclude that about two-thirds of the stars in the Galaxy formed in an early collapse phase during which the metal abundances in the interstellar gas - well mixed by turbulence - rose from zero to between a half and three-quarters of the solar value. Following the collapse to the galactic disc, ordered motion was established and no further large-scale mixing took place, the metal abundance rising in some places to 1.5 times that in the Sun, in other places remaining at a half.

Further investigations appear to confirm Dixon's analysis. Powell (1970) has found differences in iron abundances, among 12 F-dwarfs, by factors up to 3:1, and Nissen (1970a) has found differences up to 8:1 among 53 F5-G2 stars. However, the interpretation of observational data is much in dispute. Whereas it is argued that the metal abundances in a young star cluster in Sculptor are less than in the Hyades by a factor 5 (Eggen, 1970) and that the old clusters M67 (age 3×10^9 yr) and NGC 188 (age 7×10^9 yr) both have metal abundances greater than the Hyades by a factor of 2 to 5 but probably without any corresponding increase in helium abundance (Demarque and Schlesinger, 1969; Aizenman et al, 1969; Eggen and Sandage, 1969; Spinrad and Taylor, 1969; Spinrad et al, 1970; Torres-Peimbert, 1971; Torres-Peimbert and Spinrad, 1971), others conclude, sometimes from the same data, that the abundances are not greater than in the Sun (Bessell, 1972; Barry and Cromwell, 1974; Pagel, 1974). Metal abundances up to four times the solar value may be common even among stars as old as or older than the Sun. Nevertheless, superimposed on these local variations there appears to be an overall general increase in heavy element abundances with time, Fig. 7.06 (Pagel, 1970b; Hearnshaw, 1972; Powell, 1972; Clegg and Bell, 1973). So far as the time scale is concerned, Bond (1970) concludes that the initial collapse phase of the Galaxy must have occurred before the synthesis of 90% of the heavy element content now in Population I stars, since for stars with more than one tenth of the solar metal abundance there is little or no correlation between galactic orbital eccentricity and chemical composition.

Powell (loc. cit.) shows data indicating $d \log Z / d \log t \approx 1$. The data of Clegg and Bell (loc. cit.) indicates that heavy element abundances have increased by a factor six during the last 90% of the life of the Galaxy and three-quarters of the stars were formed during the last 4×10^9 yr with heavy element abundances greater than or equal to the solar value.

In addition to the local and apparently random variations in stellar element abundances over short distances described above, there appear to be systematic changes with location on the large-scale within galaxies, and from galaxy to galaxy.

Cepheid variables in the Sagittarius arm appear to be richer in metals than those in the solar region, whereas in the Cygnus arm they appear to be metal deficient (Williams, 1966). Evidence of metal abundances in old stars in the general field and in globular clusters indicates that there is a general tendency towards decreasing average metal abundances with increasing distance from the centre. Within 200 pc of the centre the abundances are greater than in the Sun; between 200 pc and 500 pc about equal to the Sun, and beyond 1 kpc from the centre they decline to a half to a fifth the solar abundances. Care must be taken in interpreting such variations - they could be due to, say, dependence of metal abundance on the rate or mass spectrum of star formation (stars which synthesise and eject the metals) with these latter parameters varying with position, so that locally they would be regarded as time-dependent variations in metal abundance.

The metal abundances of globular clusters of stars in the Andromeda galaxy are on average similar to the more metal rich globular clusters in our Galaxy; the metal deficient clusters which are most common in the Galaxy are rare in Andromeda (van den Bergh, 1969a). The centre of the disc of M31 appears to be metal rich (at least as rich as the galactic cluster M67) (Spinrad and Taylor, 1971). Between 200 pc and 500 pc from the centre (and at 1 kpc from the centre of NGC 4472) the abundance is solar, and farther out it is weak-lined but not extreme Population II as envisaged by Baade, (Spinrad et al, 1971). M32, the dwarf elliptical companion to M31, has a mildly metal poor globular cluster type of spectrum (Spinrad and Taylor, loc. cit.).

Analysis of the stellar contents of galaxies indicates that stars in galactic nuclei are generally metal rich (Spinrad, 1970; Welch and Forrester, 1972); as in the case of the CNO group of elements in the interstellar gas (section 7.073) the abundances of metals in the stars probably decreases by about an order of magnitude from the centre of the galaxy towards the edge.

Spinrad has shown that stars in the outer parts of our Galaxy, including the globular clusters, are metal deficient in comparison with the dwarf E_p galaxy NGC 205, which is in turn metal deficient in comparison with the nucleus of M31, whereas the centre of the Galaxy appears to be metal rich (van den Bergh, 1972). Metal abundances of stars in galaxies also appear to increase with the total mass of the galaxy (Spinrad and Peimbert, 1970). The SMC appears to be deficient in both heavy elements (Przybylski, 1972; Robertson, 1973) and in dust (MacGillivray, 1975).

The picture is therefore somewhat confused, but it may perhaps be summarised as follows.

In any large region in a galaxy, heavy element abundances have increased with time. In the early stages (say the first 10^9 yr or thereabouts) the interstellar gas is well mixed and there are no local abundance variations. Thereafter, the cessation of efficient mixing and local enrichment results in abundance differences by a factor two or more developing over distances of only ~ 100 pc. Over a prolonged period of time enrichment continues more rapidly towards the centres of massive galaxies than towards their edges or in dwarf galaxies.

So far as abundance variations from element to element are concerned, perhaps the most important general trend is that, in stars which are relatively deficient in heavy elements, the deficiency increases along the sequence of elements produced by the α, e, s(light) and s(heavy) processes, Fig. 7.07, lending support to the view that these elements are synthesised in stars which explode.

The difficulties involved in the interpretation of the observational data have been summarised by Unsöld (1969); they are considerable and leave any attempts to draw more detailed conclusions at the present time open to doubt.

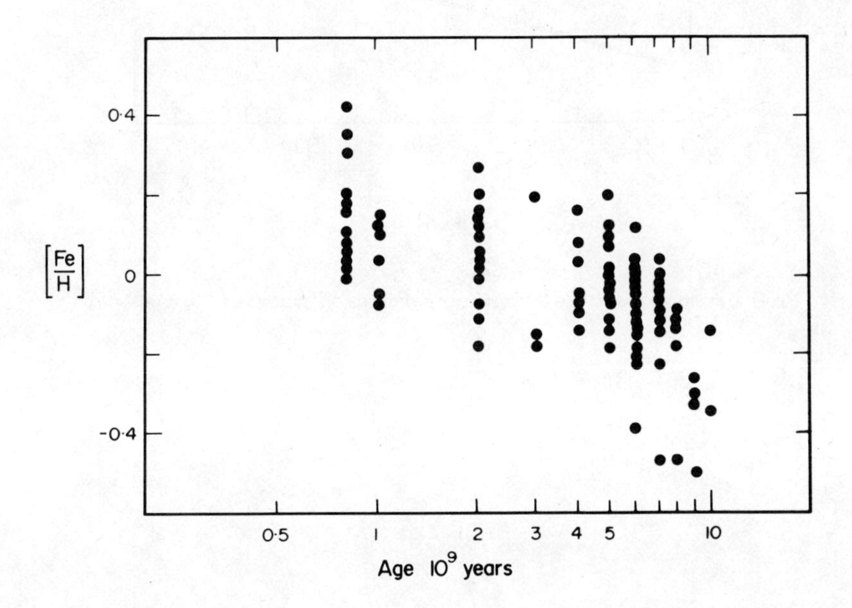

Fig. 7.06
Correlation of ages of nearby F stars with metal abundance deduced from ultraviolet excess.

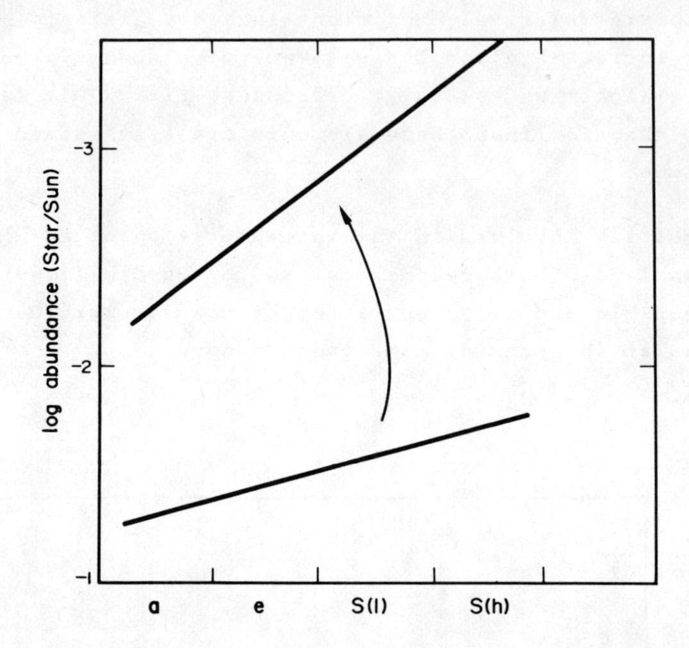

Fig. 7.07
The dependence of element abundance on mode of formation.

PART 2: THEORY

Chapter 8.

CLOUD STRUCTURE IN THE INTERSTELLAR GAS.

The cloud-like structure of the interstellar gas is the most apparent feature of its distribution. It was noted in chapters 2 and 7 that the ratio of the density in a cloud to that in the intercloud gas may range from as little as two to as much as 10^5 or possibly 10^{10}. The high density clouds are found predominantly in regions of star formation. To begin our study of how stars form from the interstellar gas, we should consider how a cloud structure is maintained and how it is initiated.

8.01 The Maintenance of a Cloud Structure

If the cloud structure is stable, there must be pressure equilibrium between the cloud and the surrounding gas; if it is not stable, clouds must be reformed continuously.

Pressure equilibrium implies that the pressure must have the same value at two or more densities - those of the cloud and of the surrounding gas - and therefore that the pressure density relation, to be continuous, must have a form similar to that shown on Fig. 8.01.

At A and C on Fig. 8.01, the pressure increases with the density. Consequently at those points, if the gas is compressed to higher density, the pressure within the compressed region increases and acts to re-expand the gas. If the gas is expanded the internal pressure falls and the gas is re-compressed. A and C are therefore stable conditions. The opposite is true at B; compression of the gas leads to reduced pressure inside the compressed region and further compression by the external pressure, so that gas compressed at B will change density until it reaches equilibrium at C. Similarly gas expanded at B will reach equilibrium at A.

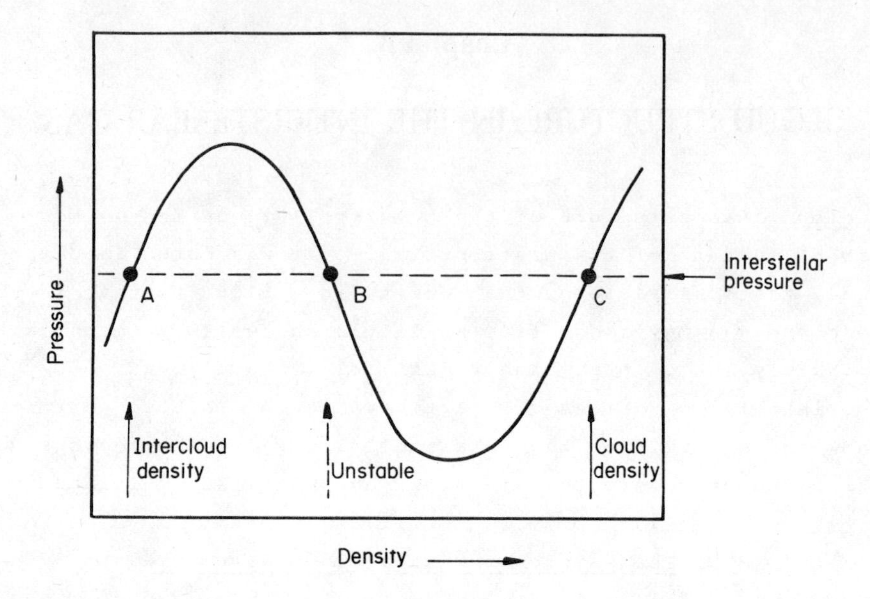

Fig. 8.01
Pressure density relation necessary for an equilibrium cloud structure in the interstellar gas.

The implications of Fig. 8.01 are seen in terms of the gas equation

$$\left.\begin{aligned} p &= nkT \\ \rho &= nm \end{aligned}\right\} \qquad (8.01)$$

Bearing in mind that most of the gas is hydrogen, the average mass per particle, m, may decrease by a factor 2 due to ionisation or increase by the same factor due to combination into H_2 molecules. It was pointed out earlier, however, that disparities between the cloud and intercloud densities range from a factor of two upwards to five or ten orders of magnitude and these must be due to differences in n. Pressure equilibrium requires corresponding differences in T and therefore in Fig. 8.01 for constant m,

$$\frac{T_A}{T_C} = \frac{\rho_C}{\rho_A} >> 1 \qquad (8.02)$$

The cloud temperature is determined by the balance of heating and cooling, and the problem of the maintenance of a cloud structure in the interstellar gas therefore reduces to the question "Does the dependence on density of the heating and cooling of the gas give a relationship of the form of Fig. 8.01?". The study of this problem is due largely to Field (1962), Pikelner (1968) and Field et al, (1969).

8.011 Heating of the Interstellar Gas

A major source of heating of the interstellar gas is believed to result from collisional ionisation by low energy cosmic rays (Hayakawa et al, 1960). Interstellar X-rays may also contribute substantially.

Low energy charged particles in the cosmic rays do not reach the earth and therefore cannot be detected directly. Estimates of the energy flux contained in them are indirect and differ considerably. For instance, Pikelner (loc. cit.) derives 10^{-12} erg cm^{-3} in 25 MeV protons as being necessary to give the observed interstellar ionisation, whereas Field et al, (loc. cit.) estimate 6×10^{-14} erg cm^{-3} in 2 MeV protons.

Protons with energies in the range given are non-relativistic.

Let H denote the rate of heating of the gas, I the rate of ionisation and K the kinetic energy of each ejected electron; then

$$H = I.K \qquad (8.03)$$

Let $\sigma(E)$ be the collision cross section of a cosmic ray proton (CR) with a hydrogen atom. The number of collisions suffered by the cosmic ray passing through a cloud of hydrogen atoms of density n_H is then given by

$$\text{Collision rate} = n_H\,\sigma(E)\,v(E) \qquad (CR)^{-1}\,s^{-1}$$

$$= n_H\,\sigma(E) \qquad (CR)^{-1} \text{ per unit length}$$

$$= \sum_E n_H\,\sigma(E)\,N(E) \qquad vol^{-1}\,s^{-1} \qquad (8.04)$$

where N(E) is the flux of cosmic ray protons. Each collision of a CR proton with a hydrogen atom will eject one electron, so that the rate of ionisation is equal to the collision rate: thus

$$I = \sum_E n_H\,\sigma(E)\,N(E) \qquad vol^{-1}\,s^{-1} \qquad (8.05)$$

Let the energy lost by a CR per unit mass of gas traversed (mostly hydrogen) be denoted $\varepsilon(E)$. The kinetic energy per electron ejected is then

$$K = m_H\,\varepsilon(E) - X \qquad (8.06)$$

where X is the ionisation potential. This is subtracted because it is radiated away on recombination. Equations (8.03), (8.05) and (8.06) then give for the rate of heating of the gas by cosmic rays

$$H = n_H \sum_E \sigma(E)\,N(E)\,\{m_H\,\varepsilon(E) - X\} \qquad (8.07)$$

Note that <u>the heating rate is proportional to the gas density n_H.</u>

Using for the CR energy density the value suggested by Pikelner, 10^{-12} erg cm^{-3} in 25 MeV protons gives $\Sigma N(E) = 170$ cm^{-2} s^{-1}. Energy losses per ionisation are about 35 ev in air and vary little with the CR energy: X is 13.6 ev for hydrogen and the net energy loss for ionisation will therefore be ~ 20 ev. With $\sigma \sim 10^{-17}$ cm^2 this gives an ionisation rate of 1.7×10^{-15} s^{-1} per hydrogen atom and a heating rate

$$H \approx 5.10^{-26} \, n_H \text{ erg cm}^{-3} \text{ s}^{-1} \qquad (8.08)$$

The relationship is shown in Fig. 8.02.

Although this account is greatly simplified, the effect of the approximation is much less than the uncertainty in the flux of low energy cosmic rays.

Several estimates have been made of the ionisation rate which may be compared with that derived above namely, 1.7×10^{-15} s^{-1} H atom^{-1}. Hjellming et al, (1969) deduce $(2.5 \pm 0.5) \times 10^{-15}$ s^{-1} per H atom from electron densities derived from analysis of radio frequency observations of pulsars. On the other hand, Field et al, (1969) similarly deduce a value 4×10^{-16} s^{-1} per H atom from a variety of data.

It was pointed out above that X-rays may also be a significant source of heat in the interstellar gas. Several difficulties associated with cosmic ray heating were listed by Werner et al, (1970). The assumed existence of a large flux of low energy cosmic rays rests on an enormous extrapolation of data at higher energies. An upper limit of 10^{-16} s^{-1} on the rate of ionisation due to cosmic rays is set both by the observed limit on the abundance of spallation products which would be created in the collisions, and by the observed abundance of H_2 which puts a limit on the rate of dissociation of H_2. The latter problem is not so acute in the case of X-rays because unlike the cosmic rays they cannot penetrate the denser clouds.

It appears also that, if cosmic rays are produced locally in supernovae explosions, they will produce a local heating since they cannot diffuse fast enough to produce uniform heating throughout the interstellar gas.

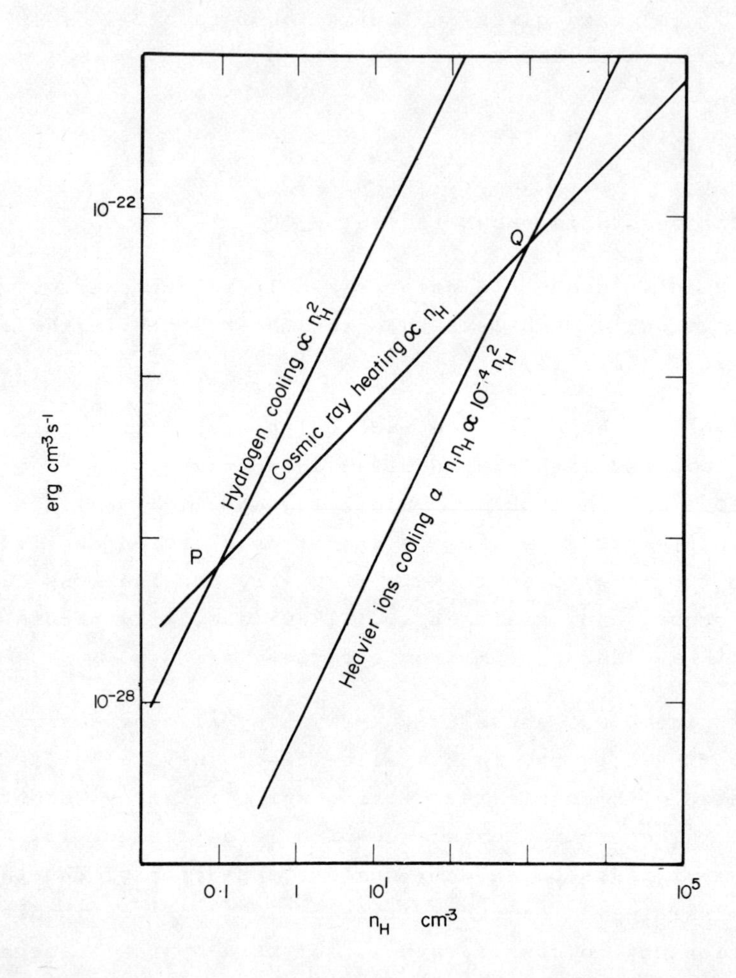

Fig. 8.02
The dependence on gas density of the rates of heating and cooling.

Two significant differences arise if heating is by X-rays rather than by cosmic rays. Firstly, ionisation of helium contributes substantially to the electron density because the photo-ionisation cross section of helium is twenty times that of hydrogen, and helium recombinations produce photons able to ionise hydrogen. Secondly, at high electron densities $n_e/n_H \geqslant 0.1$ cosmic rays lose much of their energy in collisions with electrons. Only some of the energy thus transferred produces ionisations by collisions of the more energetic electrons with atoms, whereas X-rays continue to interact directly with the atoms to produce more ionisation. That is to say, in comparison with cosmic ray heating, when $n_e/n_H \geqslant 0.1$ X-rays produce at one and the same time a higher electron density and a lower electron temperature.

Excepting these differences, heating by X-rays is similar to that by cosmic rays, taking the form outlined in the preceding equations. A detailed study is given by Habing and Goldsmith (1971).

It should be remembered that, as in the case of heating by cosmic rays, the assumption of a significant flux of low energy X-rays depends on a large extrapolation from high energies and requires the flux to be an order of magnitude greater than such an extrapolation indicates.

The latest observational data lead to conflicting conclusions. Recent estimates of the ionisation rate in the interstellar gas based on ultraviolet spectrometry range from 8×10^{-15} s^{-1} (Penston, 1974) to $< 4 \times 10^{-17}$ s^{-1} (Meszaros, 1974). Although the flux of X-rays with energies > 0.2 keV gives an ionisation rate too low by a factor twenty (Ilovaisky and Ryter, 1971), the X-ray flux is rising steeply below 0.28 keV (Yentis et al, 1972). On the other hand, there are now several indications that the ultraviolet flux from OB stars may be the dominant cause of heating of the interstellar gas (Walmsley and Grewing, 1971; Hills, 1972; Castle and Hills, 1974; Torres-Peimbert, et al, 1974). The various kinds of data and the analyses that can be made of them have been reviewed by Steigman et al, (1974).

8.012 Cooling of the Interstellar Gas

Cooling of the interstellar gas occurs as a result of collisional excitation of atoms and ions by thermal electrons, the energy being radiated away. At high temperatures (several thousand K) excitation of hydrogen and helium predominates, whereas at low temperatures ions such as C^+, Si^+, Fe^+ are the main agents.

Let Q represent the cross section for excitation by electrons of an atom or ion which occurs with number density n. Then:

$$\text{excitation rate per electron} = v_e Qn$$

where v_e is the electron velocity, and

$$\text{excitation rate per unit volume} = v_e Qnn_e$$

Now:

$$v_e = \sqrt{(3kT/m_e)}$$

and

$$Q \propto b \,.\, \text{oscillator strength}/\text{kinetic energy of electron.excitation energy W}$$

$$= b \,.\, f/\tfrac{3}{2}kT.W$$

where b is the guillotine factor: $b(T) = 0$ for $\frac{3}{2}kT < W$, when the electron has insufficient energy to excite the atom or ion. Denoting the rate of cooling per unit volume by H' we then have for each kind of atom or ion:

$$H' = (\text{rate of excitation/unit volume}).(\text{excitation energy W})$$

$$= n_e n \sqrt{(3kT/m_e)}.fWb/\tfrac{3}{2}kTW$$

$$= \alpha bfT^{-\frac{1}{2}} n_e n \qquad (8.09)$$

where $\alpha = 2/\sqrt{(3km_e)}$.

n_e is determined largely by the ionisation of hydrogen by cosmic rays and it does not vary with the temperature of the interstellar gas until the temperature becomes high enough for ionisation of hydrogen by thermal electrons, ~ 10^4 K. Therefore $n_e = \beta n_H$ where β is generally $\approx 10^{-1}$ except at temperatures $\gtrsim 10^4$ K when $\beta \simeq 1$ as a result of almost complete thermal ionisation. Consequently at temperatures high enough to excite but not to ionise the hydrogen, $n = n_H$ and

$$H'(T_{high}) = \alpha\beta b_H f_H T^{-\frac{1}{2}} n_H^{\,2} \qquad (8.10)$$

As T falls below several thousand K, b_H becomes zero and cooling by hydrogen effectively ceases, cooling by heavier ions taking over. The abundance of the ions C^+, Si^+, Fe^+ is $\simeq 10^{-4}$ n_H; i.e. $n \simeq 10^{-4}$ n_H and consequently the cooling rate at the lower temperatures is:

$$H'(T_{low}) \simeq 10^{-4} \alpha\beta b_i f_i T^{-\frac{1}{2}} n_H^{\,2} \qquad (8.11)$$

The predominant difference between the latter two equations is the factor 10^4 in the constants of proportionality due to the lower abundance of the heavier ions. Cooling therefore depends on gas density in the manner shown in Fig. 8.02.

8.013 Temperature, Density and Pressure

Let us consider cooling by hydrogen and examine what happens if the density n_H decreases, when T < 10^4 K. The cooling rate is proportional to $n_H^{\,2}$, equation (8.10) but the heating rate is proportional to n_H, equation (8.07); therefore as n_H decreases, the rate of cooling falls faster than that of heating (Fig. 8.02) and the temperature rises, Fig. 8.03, causing the cooling rate to fall faster still since it varies also with $T^{-\frac{1}{2}}$. The situation is thermally unstable and the temperature rises rapidly until it reaches ~ 10^4 K and then $\beta = n_e/n_H$ rises suddenly due to thermal ionisation of hydrogen, Fig. 8.04.

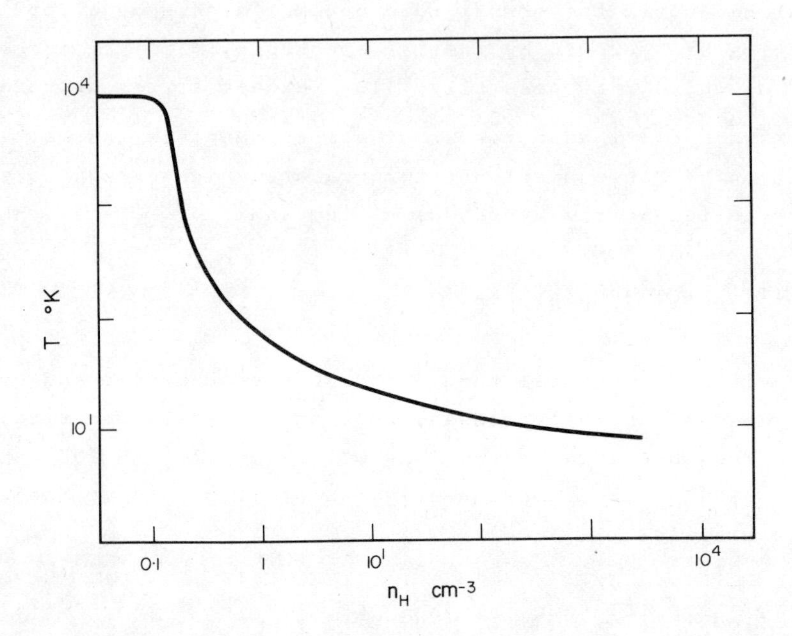

Fig. 8.03
The dependence of temperature on density in the interstellar gas, consequent upon the heating and cooling rates shown in Fig. 8.02.

This causes the cooling also to rise suddenly with a further decrease in density if necessary, until it equals the heating, point P on Fig. 8.02.

Now consider what happens when the density is increased again from point P. Cooling increases faster than heating and the temperature drops. When it has fallen to a level at which hydrogen is no longer excited, b(H) = 0 and heavier ions take over the cooling at a level ~ 10^4 below the cooling rate by hydrogen. The cooling rate is now far below the heating rate, but increases faster as n_H increases until thermal balance is

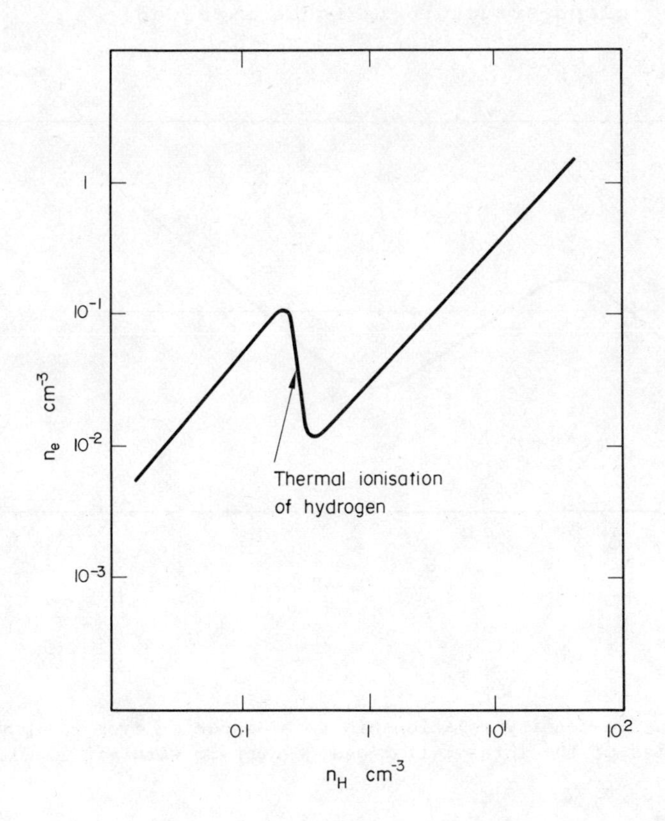

Fig. 8.04
The level of ionisation as a function of gas density, resulting from the temperature - density relationship shown in Fig. 8.03.

achieved again at Q. Of particular interest to us is the dependence of gas pressure $(n_e + n_H)$ T on density n_H; this can be determined from Figs. 8.03 and 8.04 and gives the result shown in Fig. 8.05. Reference to Fig. 8.01 and the introduction to this chapter shows that Fig. 8.05 possesses the essential features necessary for stable pressure equilibrium to exist at two very different densities, giving clouds at C embedded in an intercloud medium at A.

The physical properties of the interstellar gas are therefore such that a cloud structure is to be expected.

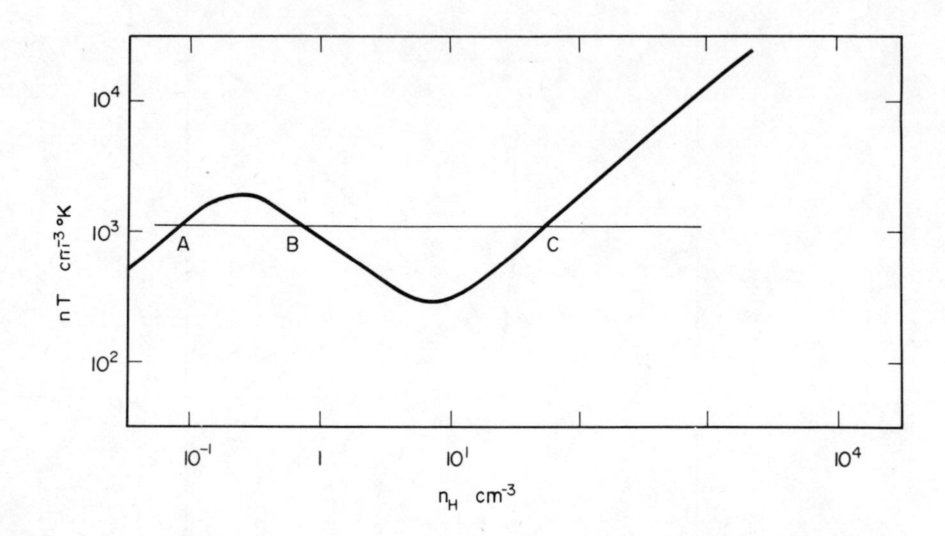

Fig. 8.05
The pressure-density relationship to be expected from the physical properties of the interstellar gas. Compare with Fig. 8.01.

8.02 The Generation of a Cloud Structure

It has been seen in the preceding sections that interstellar gas is in stable pressure equilibrium at two very different densities and that we may expect to find, as in fact we do, clouds embedded in an intercloud medium. The question arises, however, as to what forces act on the gas to distribute it between these two density states; why, for instance, is not all the gas in one state or the other?

Isotropic forces, such as cosmic ray pressure, and large-scale forces such as the dynamical effects of galactic rotation, may produce large-scale changes in gas density or motion, but the forces which are effective in producing a cloud structure must be as localised as the clouds themselves. The origin of such forces is therefore most likely to be found in the energy of the stars, and of stellar explosions.

A mechanism by which stellar energy may be transferred to the interstellar gas in a manner likely to produce clouds in the gas has been proposed by Oort (1954) and studied in some detail by Oort and Spitzer (1955), by Field and Saslaw (1965), and by Penston et al, (1969). The mechanism is imagined to operate in the following manner.

An aggregate of massive, hot stars - an O-association - forms by the condensation of the central denser part of a large massive cloud of cool interstellar gas having a temperature of 100 K or less. These stars ionise the gas in their immediate surroundings, producing a compact HII region (see chapters 2 and 7) and raising the kinetic temperature of the gas in the ionised region to 10^4 K. The pressure in this region is then a hundred or more times that in the surrounding neutral gas, and the ionised gas expands pushing the surrounding gas outwards and compressing it. Pressure equilibrium will be re-established when the density of the surrounding gas has increased about a hundredfold. Reference to Fig. 8.05 shows that such a change in density is sufficient to move gas through the unstable region around B, to C. The mechanism therefore appears to be strong enough to form clouds out of the intercloud medium; the expansion of the gas ionised by the aggregate of stars causes movement in the opposite direction on Fig. 8.05, from C to A; the Figure of course represents the initial and final states of pressure equilibrium. The speed of expansion of the HII region will be about the speed of sound in the hot gas, 17 km s^{-1} or 17 pc in 10^6 yr. Since the lifetime of the O stars will be several million years, the cool gases will be compressed and then driven some tens of parsecs away from the stars. This will be so even if the surrounding cool gases have considerable mass, because evaporation of gas from the sides of the compressed regions facing the star will produce a rocket effect, accelerating the cool cloud outwards.

Density fluctuations in the cloud in which the O-association formed will be increased by this mechanism since the ionisation will penetrate the less dense gas between more dense regions, isolating and compressing them into clouds.

The small clouds which are formed and ejected from the massive clouds in this manner suffer inelastic collisions with other similar small clouds ejected from other aggregates, coalescing to build up new massive cloud complexes. In time these become so massive that they contract under self gravitation, aggregates of O-stars form in their centres, and the process repeats itself. Thus there is a feedback process:-

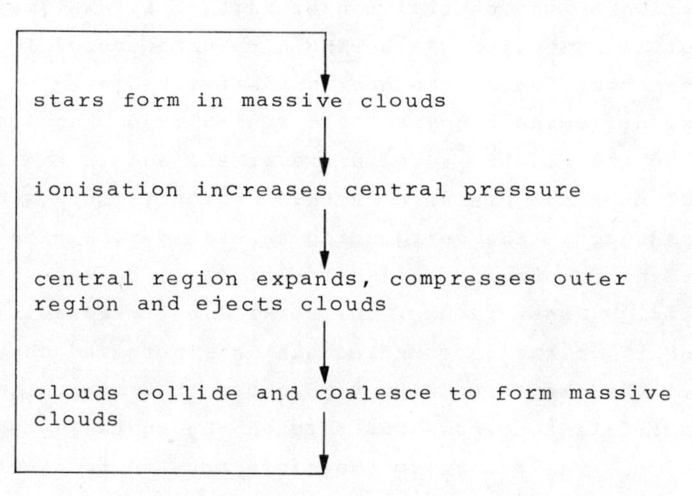

The numerical studies of this process are concerned with two features of it - how long it takes to get through the stages in the process, and how much mass is left at the end of each cycle.

For a massive cloud in which the force of gravitational self-attraction greatly exceeds the internal gas pressure, the contraction time (see chapter 10) is given by the free fall collapse time

$$t = (4\pi G \rho)^{-\frac{1}{2}} \approx 3 \times 10^7 \, n_H^{-\frac{1}{2}} \text{ yr.} \tag{8.12}$$

Densities in large clouds (chapters 5 and 7) are typically $\geqslant 10^{-23}$ g cm^{-3}, for which $t \leqslant 10^{7}$ yr, and this may be taken to indicate the time for stars to form in massive clouds - although the lack of a definite understanding of the detailed processes by which stars form from clouds must cause us to regard this assumption with caution.

When the O-stars form in the cloud, the ionisation front moves outwards until it encloses a volume within which ionisation of hydrogen atoms by absorption of Lyman continuum photons balances recombination of electrons and protons. That is to say, all Lyman continuum photons from the stars are absorbed to re-ionise recombined atoms. The time for the ionisation front to reach the limit - the radius of the Strömgren sphere - will be about the time required for the stars to emit as many Lyman continuum photons as there are H atoms within the Strömgren sphere, since once the ionised region is set up the fraction of neutral atoms within it at any time is small and their number density must be low in comparison with the photon flux, except towards the outer edge.

For example, consider the growth of an HII region around an O5 star having an effective temperature of 35,000 K and a luminosity of 3 x 10^5 suns. The flux of radiation from the star is

$$L = 3 \times 10^{5} L_{\odot} = 1.2 \times 10^{39} \text{ erg s}^{-1} \qquad (8.13)$$

Representing the energy distribution by black body radiation, the fraction of this which occurs at wavelengths shorter than the limit of the Lyman continuum is given by the Planck function (Allen, 1964)

$$F_{o-\lambda} / F_{total} = 0.30 \qquad (8.14)$$

The peak energy flux occurs at λ_m where by Wiens' Law

$$\lambda_m T = 0.29 \text{ cm deg} \tag{8.15}$$

and so

$$\lambda_m = 830 \text{ Å} \tag{8.16}$$

This is in the Lyman continuum and corresponds to photons having energies of 14eV = 2.2×10^{-11} erg. Consequently the rate of emission of ionising photons is given approximately by

$$L_i \approx (0.30) \times (1.2 \times 10^{39})/2.2 \times 10^{-11}$$
$$= 1.6 \times 10^{49} \text{ s}^{-1} \tag{8.17}$$

In equilibrium this will be equal to the rate of recombination of electrons and hydrogen ions in the HII region. The rate of recombination per unit volume is

$$\sigma v_e n_e n_i \tag{8.18}$$

where σ is the recombination cross section, v_e the thermal velocity of the electrons and n_e, n_i the number densities of electrons and ions. When ionisation is virtually complete, $n_e = n_i = n_H$, the number density of hydrogen atoms prior to ionisation. If r is the radius of the HII region then writing $\sqrt{(3kT/m_e)}$ for v_e we have

$$L_i = \frac{4}{3}\pi r^3 \sigma \sqrt{(3kT/m_e)}\ n_H{}^2 \tag{8.19}$$

Hence the radius of the HII region is

$$r = S_o n_H{}^{-2/3} \tag{8.20}$$

where

$$S_o = (3L_i m_e^{\frac{1}{2}}/4\pi\sigma\sqrt{(3kT)})^{1/3}$$

For a more comprehensive formulation the reader is referred to Seaton (1960). It should be borne in mind that although the formulation given here is approximate, in any given situation the uncertainty in L_i is considerable.

Now $T \approx 10{,}000$ K and at that temperature $\sigma = 9 \times 10^{-21}$ cm^2 (Allen, 1964).

Consequently

$$S_o = 2.8 \times 10^{-15} L_i^{1/3} \text{ pc} \qquad (8.21)$$

and in the example given with L_i from equation (8.13) this gives $S_o = 70$ pc. The radius r of the HII region is given by equation (8.20) and the total number of hydrogen atoms in this volume prior to ionisation is

$$N = \frac{4}{3} \pi S_o^3 n_H \qquad (8.22)$$

With $S_o = 70$ pc this gives

$$N = 3.9 \times 10^{61} n_H \qquad (8.23)$$

The time taken to ionise these atoms is the time taken to emit N ionising photons at a rate L_i; thus

$$t(\text{ionisation}) = N/L_i$$

$$= 8 \times 10^4 n_H^{-1} \text{ yr} \qquad (8.24)$$

Since $n_H \geqslant 10$ cm^{-3}, this is much less than the gravitational collapse time given by equation (8.12). That is to say, the second stage in the feedback process, the increase in the central pressure due to ionisation, takes place in a time short in comparison with the first stage - the gravitational collapse of the cloud from which the stars form.

The third stage in the process is the expansion of the ionised region. This expansion will occur with the speed at which a pressure wave moves outwards through the ionised region - the speed of sound, given by

$$\begin{aligned} V &= (\gamma p/\rho)^{\frac{1}{2}} \\ &= (\gamma RT/\mu)^{\frac{1}{2}} \end{aligned} \tag{8.25}$$

using the perfect gas equation, where γ is the ratio of specific heats at constant pressure and at constant volume, R is the gas constant and μ the molecular weight. In completely ionised hydrogen where no energy is being used to ionise the gas, $\gamma = 5/3$; the molecular weight $\mu = \frac{1}{2}$, and $T \approx 10^4$ K; hence

$$V = 17 \text{ km s}^{-1} \tag{8.26}$$

The time to double the size of the HII region will, using equation (8.20) be

$$t = S_o\, n_H^{-2/3}/V \tag{8.27}$$

and taking again the example used above, with $S_o = 70$ pc and $V = 17$ km s^{-1} we have

$$t(\text{expansion}) = 4 \times 10^6\, n_H^{-2/3} \text{ yr} \tag{8.28}$$

which with $n_H = 10$ cm^{-3} gives

$$t(\text{expansion}) \approx 9 \times 10^5 \text{ yr}$$

This is comparable with but rather less than the gravitational collapse time given by equation (8.12).

The fourth and final stage in the feedback process is the collision and coalescence of small clouds building up the large complexes to start the process again. An estimate of the time involved can be made by considering a simple model. Suppose that there is initially a distribution of small clouds, and assume that collisions between them are inelastic. Let there be N clouds per unit volume, each cloud having mass m, radius r and velocity v. The time interval between collisions is

$$t_o = (4\pi r^2 N v)^{-1} \qquad (8.29)$$

At the end of this interval coalescence will have produced a distribution of cloud masses, but a simple representation of the situation is that, on average, clouds have coalesced in pairs to produce N/2 clouds per unit volume each with mass 2 m and radius $2^{2/3}r$. Assuming for the moment that the velocities are unchanged, the time interval between collisions is now

$$t_1 = 2^{1/3}\, t_o \qquad (8.30)$$

and the process repeats. That is to say, each time the mass doubles, the time interval for doubling increases by $2^{1/3}$. Thus we have

$$m = m_o \quad \text{at} \quad t = 0$$

$$m = 2^n m_o \quad \text{at} \quad t = t_o \sum_o^n 2^{n/3} \qquad (8.31)$$

where $\sum_o^n 2^{n/3} = (2^{(n+1)/3} - 1)/(2^{1/3} - 1)$

and t_o is given by equation (8.29).

The observations by Heiles of HI 'cloudlets', see chapters 2, 5 and 7, provide the relevant initial parameters:-

$r = 10^{19}$ cm

$v = 5 \times 10^{5}$ cm s^{-1}

$N = 5 \times 10^{-60}$ cm^{-3}

$m_o = m_\odot$

These give $t_o = 10^7$ yr. To build up large complexes with masses of about 10^5 suns requires $n \simeq 17$ and a time $t = 2.4 \times 10^9$ yr. This is similar to the time scale for star formation in the Galaxy and conflicts with the conclusion of Field and Saslaw who assumed the time scale for cloud build-up to be of the order of $5t_o$. They concluded that since this time scale is only about one thousandth of the age of the Galaxy, only about one thousandth of the mass of a complex must condense into stars at each cycle, the remainder being ejected back into the interstellar medium by the Oort process as small clouds. This conclusion is not supported by the analysis presented here; but it must be borne in mind that as yet we know little of the distribution of forces acting on clouds and it may be that they are gathered together much more effectively than is suggested, perhaps being swept up by the advancing density wave on the leading edge of spiral arms as discussed in chapter 9.

It would be valuable to have measurements of the total mass of HI as well as of HII connected with compact HII regions, to compare with the estimated masses of the aggregates of stars which excite the nebulae.

Existing evidence is conflicting. von Hoerner (1968) calculates the mass of the cloud aggregate from which an average association of OB stars forms and concludes that it is 3×10^5 $M_\odot$ and that only one per cent of the cloud condenses into stars. The calculation is, however, based on the assumption that the velocity dispersion in the cloud before the stars form is the same as the velocity dispersion of the stars formed, $\approx$ 5 km s^{-1}. This does not appear to be supported by measurements of velocity dispersions in cold clouds reported in chapter 7 which indicate velocities of a few tenths of a km s^{-1} at the most. Since the calculated mass depends on v^2 it is reduced by a factor ~ 100 and leads to the conclusion that most or all the mass condenses into stars. Mathews (1969) reaches a similar conclusion from an analysis of the compact HII region W49; he calculates that the total stellar mass is of the order of a hundred times the mass of gas in the surrounding HII region. This does not take account of additional HI in which the system may be embedded, but there are no strong HI absorption lines at 21 cm which would indicate a substantial mass of HI.

The first part of this chapter showed that the physical properties of the interstellar gas are such that a cloud-like structure is to be expected in it. The proportions of the mass of interstellar gas which are contained in the clouds and in the intercloud regions must depend on the forces of compression and expansion acting on the gas to change any particular regions from stable cloud state to the stable intercloud state (Fig. 8.01), or *vice versa*. It is reasonable to argue that these forces must be localised in order to produce a mixture of the states, and that localised forces on the gas are produced by the stars as Oort suggests. However, the Oort model shows a manner in which small clouds may be built up into large clouds, and large clouds broken down into small ones, but its effect on the division of mass between cloud and intercloud must be through the compression exerted on the intercloud medium by the material ejected from HII regions, tending to produce clouds, and the evaporation or expansion of clouds back into the intercloud medium. This requires that the gas surrounding a newly formed aggregate of stars should be acted upon by that aggregate and its residual gas, but it is not necessary that the surrounding gas should be part of the aggregate. The time scale for interaction of star formation with its environment in the manner required - by gas pressure waves - is the sum of the times for the first three of the four stages of the Oort process; that is about 10^7 yr, and since this is itself the sum of contraction and expansion times for clouds, it is evidently a reasonable time scale for processes which change the cloud densities from one stable state to another. It may be concluded that star formation produces localised pressure changes of the amounts and in the times required to produce density variations, leading to a two component cloud-like interstellar gas as discussed in the first part of the chapter.

This conclusion begs the question as to whether or not the protogalaxy gas had a cloudy structure before the first stars formed in it. It is presumed that it had, but studies of this problem, (e.g. Callebaut, 1967; Pikelner, 1970; Bird, 1968; Peebles, 1969; Weinberg, 1971; Schwarz et al, 1972; Sato, 1974) are concerned with how density perturbations grow, or how primordial clouds survive, rather than with how the density fluctuations originate - a question that we are probably unable to answer without an understanding of the origin of matter and the evolution of the Cosmos as a whole.

Chapter 9.

THE ORDERING OF CLOUDS INTO SPIRAL ARMS.

Spiral arms are the most obvious structural feature in a majority of galaxies viewed at optical wavelengths. The observational evidence summarised in the first part of this book indicates that the arms are composed essentially of clouds of gas and recently formed stars. Only young stars - OB and T Tauri stars, Wolf-Rayet stars and Cepheid variables, exhibit a spiral arm distribution. Photographs of spiral galaxies taken in the red light of old stars, and the generality of stars in our own Galaxy, show no such distribution.

The concentration of young stars into spiral arms appears to occur only because they have recently formed from the dense clouds. As the stars grow older they disperse into the general field, but the clouds do not. It is concluded that the forces which gather material into spiral arms act only or predominantly on clouds rather than on stars.

Possible forces may be divided into two kinds - gravitation and pressure forces. The latter include pressures due to the intercloud gas, electromagnetic radiation, cosmic rays and magnetic fields.

The gathering together of material into spiral arms by the action of pressure forces was for long believed to be their origin but suffered from a difficulty in that differential rotation would produce more turns on the arms over the life of a galaxy than are observed. Therefore it was necessary to suppose that the arms dissipated and re-formed at intervals $\sim 10^9$ yr. However, all massive galaxies having more than about 10^{-3} of their mass in interstellar gas appear to have spiral arms; irregulars are generally much less massive: and so there is no observational evidence to indicate periods in the lifetime of a galaxy when its arms are dissipated and renewed.

It became necessary to seek an explanation in terms of a spiral arm structure which does not wholly share the differential rotation of the mass of gas and young stars, and this led to the study of spiral density waves imposing a moving pattern on the material.

Some doubt is now cast on the observational data of differential rotation in many galaxies due to the wide discrepancies between the rotation curves derived by optical and by radio (21 cm) radial velocity measurements, the latter often showing essentially solid body rotation (Vorontsov-Velyaminov, 1969). However, there are powerful arguments that the spiral arms cannot generally be material arms wrapped into spirals by differential rotation (ibid). Most spiral arms have a fairly constant pitch angle which would require a uniform differential velocity gradient which is not observed; some galaxies have different curvatures for each of the arms in a pair, which would require two different rotation laws simultaneously in the same galaxy; and some have arms curving in opposite directions.

These various difficulties relating to the wrapping of material spiral arms by differential rotation appear to have been overcome in recent years by the studies of density waves. These studies are concerned with the effects of perturbations of differentially rotating self-gravitating discs of matter. Density perturbations of a uniformly rotating disc produce concentric rings, not spiral patterns (Hunter, 1963). Lindblad studied disc-shaped distributions of discrete mass points and Lin and Shu (1964) have studied continuous distributions of matter from the point of view of cooperative phenomena of a kind similar to those in plasma physics.

To begin with we should consider why the gas clouds are gathered into spiral arms, but the stars (excepting those recently formed from the clouds) are not. If spiral arms are due to gravitational interactions, why do these affect the clouds differently to the stars? The answer probably is that the velocity dispersions are different, being considerably greater for old stars than for young stars and clouds (chapter 7). For perturbations to affect the major structure of the disc they must have wavelengths greater than the thickness of the disc, (Toomre, 1964). The thickness of the disc increases with random motion and consequently the wavelength of effective perturbations increases with random motion. Perturbations in the disc component of gas and young stars therefore have the shortest wavelengths.

Stable perturbations do not generate spiral structure (the anti-spiral theorem; Lynden-Bell and Ostriker, 1967), but energy dissipation makes a system unstable - a 'cold' system is unstable (Toomre, loc. cit.).

9.01 The Maintenance of Spiral Density Waves

Consider, then, a circularly symmetric self-gravitating thin disc of matter in differential rotation in which individual mass points are following circular orbits about the centre. Now let us introduce a circularly assymetric perturbation by gathering the mass more closely together along one radius vector - Fig. 9.01a. The gravitational potential ϕ around a circle at radius r will now be of the form shown in Fig. 9.01b. Imagine a number of the mass points which, in the absence of the perturbation, would have the circle as their orbit and would be equally spaced around it, and suppose that they are overtaking the perturbation. The effect of the perturbation becomes evident on Fig. 9.01b; going down into the potential well the mass points speed up and therefore spread out, going up out of it they slow down and bunch together. The result is that the density concentration has now moved further around in the direction of motion of the stars - but at a lesser speed. The mass points are overtaking their own perturbation. A density wave is being maintained which, seen from a frame of reference rotating with the mass, is moving anticlockwise.

In Fig. 9.01a the perturbation has been illustrated by a radial bar, but this may not be realistic and the perturbation which

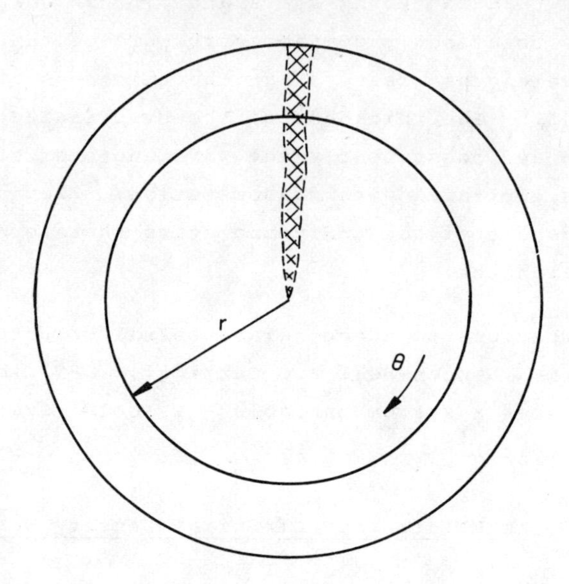

Fig. 9.01a
An increase in density along a radius in a circularly symmetric self-gravitating thin disc of matter in differential rotation, a kind of perturbation which could be the beginning of a spiral density wave.

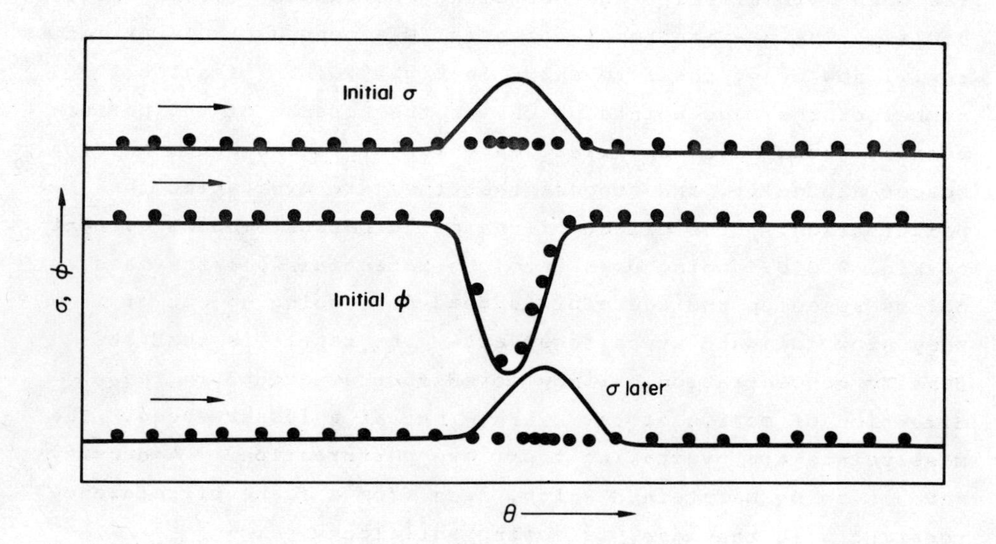

Fig. 9.01b
The initial perturbation of density and gravitational potential produce a travelling density wave.

will be of most interest to us is that which is most stable and hence longest lasting.

The density wave we are considering is an example of cooperative effects such as occur in plasma physics. Cooperation requires interaction, and the density wave which will be most stable will be that in which interactions between mass points are strongest along the wavefront. Consider mass points moving around two circular orbits of slightly different radius, and perturb the inner orbit so that it becomes an ellipse (Fig. 9.02a). There will be two regions symmetrically on each side of the major axis of the ellipse where gravitational interaction between mass points on the two orbits will be greatest - determined by their closeness of approach and similarity of motion. Thus a perturbation will spread and be maintained most strongly not along a radius vector but at an angle to it, an angle which will change only slowly because the properties of the orbits change only slowly with distance from the centre. The wavefront will therefore curve outwards and inwards from a perturbation somewhat in the form of a pair of spirals (Fig. 9.02b).

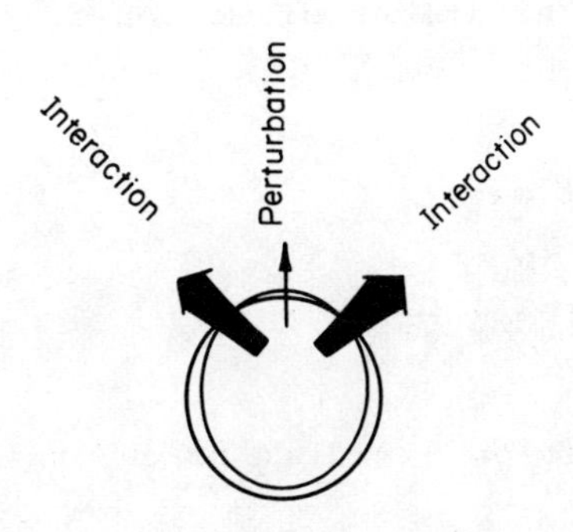

Fig. 9.02a
A circular ring of mass points perturbed to an ellipse interact most strongly with an unperturbed ring near points of intersection.

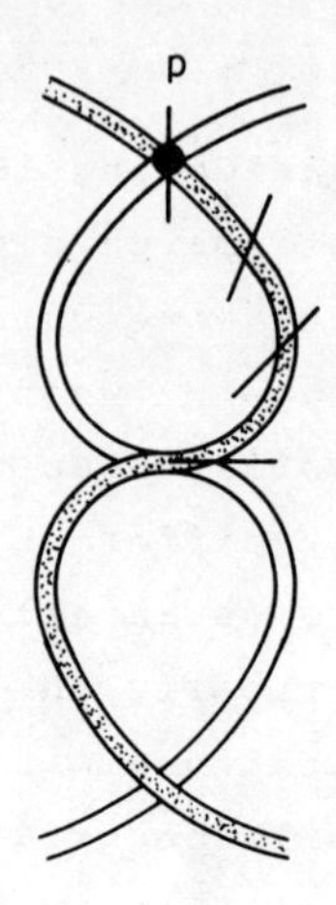

Fig. 9.02b
Interactions such as on Fig. 9.02a spread and are maintained most strongly along a spiral path.

These cooperative effects, producing spiralling density waves, have been studied by Lin and Shu (1964) in the following manner. Consider a circularly symmetric self-gravitating thin disc of matter in differential rotation and having thickness h and surface density of mass σ (Fig. 9.03). At any point in the disc distant r from the centre c, denote the gravitational potential $^{-}\Phi$. The equations of motion in cylindrical co-ordinates are then:

conservation of mass;

$$\frac{\partial \sigma}{\partial t} + \frac{1}{r}\frac{\partial}{\partial r}(ru\sigma) + \frac{1}{r}\frac{\partial}{\partial \Theta}(\sigma v) = 0; \qquad (9.01)$$

radial acceleration = radial force/unit mass;

$$\frac{\partial u}{\partial t} + u\frac{\partial u}{\partial r} + \frac{v}{r}\frac{\partial u}{\partial \Theta} - \frac{v^2}{r} = \frac{\partial \Phi}{\partial r}, \qquad (9.02)$$

azimuthal acceleration = azimuthal force/unit mass;

$$\frac{\partial v}{\partial t} + u\frac{\partial v}{\partial r} + \frac{v}{r}\frac{\partial v}{\partial \Theta} + \frac{uv}{r} = \frac{1}{r}\frac{\partial \Phi}{\partial \Theta}; \qquad (9.03)$$

conservation of gravitational flux (Poisson's Law);

$$\frac{\partial^2 \Phi}{\partial r^2} + \frac{1}{r}\frac{\partial \Phi}{\partial r} + \frac{1}{r^2}\frac{\partial^2 \Phi}{\partial \Theta^2} + \frac{\partial^2 \Phi}{\partial z^2} = -\ 4\pi G\sigma(r,\Theta)h^{-1}. \qquad (9.04)$$

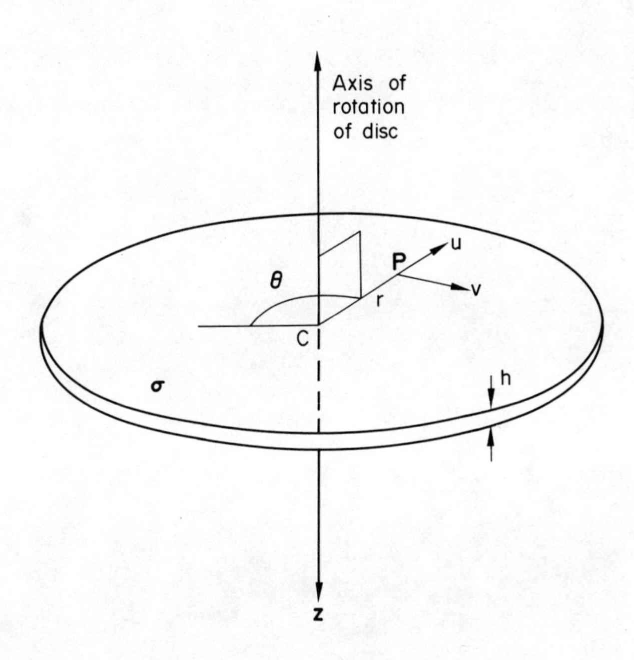

Fig. 9.03
A self-gravitating thin disc of matter. The equations of motion are given in the text.

Solutions for an equilibrium situation are of the form:

$\sigma = \sigma(r)$, circular symmetry;

$u = 0$, no expansion or contraction;

$v = v(r) = r\Omega(r) > 0$ where Ω is an angular frequency, unidirectional rotation.

Now introduce a perturbation at some point P and denote the perturbations by primes, i.e.

$\sigma = \sigma + \sigma'$

$u = u + u'$

$\Phi = \Phi + \Phi'$

etc.

Equations (9.01) to (9.04) then become, after unperturbed values have cancelled,

$$\frac{\partial \sigma'}{\partial t} + \Omega\frac{\partial \sigma'}{\partial \Theta} + \frac{1}{r}\frac{\partial}{\partial r}(r\sigma u') + \frac{\sigma}{r}\frac{\partial v'}{\partial \Theta} = 0 \qquad (9.05)$$

$$\frac{\partial u'}{\partial t} + \Omega\frac{\partial u'}{\partial \Theta} - 2\Omega v' = \frac{\partial \Phi'}{\partial r} \qquad (9.06)$$

$$\frac{\partial v'}{\partial t} + \Omega\frac{\partial v'}{\partial \Theta} + (\kappa^2/2\Omega)u' = \frac{1}{r}\frac{\partial \Phi'}{\partial \Theta} \qquad (9.07)$$

$$\frac{\partial^2 \Phi'}{\partial r^2} + \frac{1}{r}\frac{\partial \Phi'}{\partial r} + \frac{1}{r^2}\frac{\partial^2 \Phi'}{\partial \Theta^2} + \frac{\partial^2 \Phi'}{\partial z^2} = -4\pi G\sigma' h^{-1} \qquad (9.08)$$

where κ is the epicyclic angular frequency given by:

$$\kappa = 2\Omega\sqrt{(1 + (r/2\Omega)d\Omega/dr)} \qquad (9.09)$$

These equations have spiral-like solutions, e.g.

$$\Phi' = \Phi'_o(r)\exp\{in(\Omega_s t-\Theta(r)) + \Theta_o\} \tag{9.10}$$

$$\sigma' = \sigma'_o(r)\exp\{in(\Omega_s t-\Theta(r)) + \Theta_o\} \tag{9.11}$$

where:

n is the number of spiral arms, usually two;

Ω_s is the angular velocity of the spiral pattern in an inertial frame; and hence,

$\Omega_s - \Omega$ is the angular velocity of the spiral pattern in a frame rotating with the mass.

Consider now the interaction of a star in its epicyclic orbit with the passing density wave. The two will interact and reinforce the wave most strongly if the density wave passes when the star is in that part of its epicyclic orbit where its motion is most similar to that of the density wave, and it will continue to act most strongly if this situation repeats each time the spiral arm comes round. The frequency at which this occurs is, with two trailing spiral arms, (Fig. 9.04)

$$\nu = 2(\Omega - \Omega_s)/\pm\kappa \tag{9.12}$$

In a stable galaxy the numerical value of κ must lie between Ω (for solid body rotation as in a homogenous spheroidal mass distribution) and 2Ω (Keplerian orbits where all the mass is concentrated at the centre), and the maximum numerical value of ν is unity. It follows that a trailing spiral density wave will be most strongly maintained when

$$\Omega + \kappa/_2 \geqslant \Omega_s \geqslant \Omega - \kappa/_2 \tag{9.13}$$

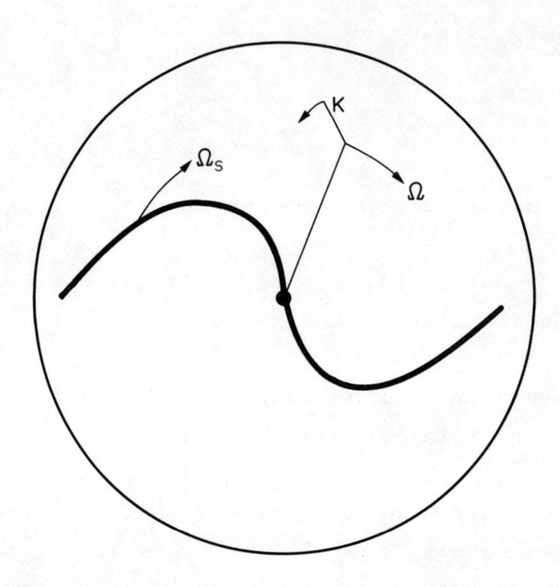

Fig. 9.04
Interaction between a star and a passing spiral density wave. See text.

The consequence of this in relation to our own Galaxy has been considered by Lin and Yuan (1969) and their result is shown on Fig. 9.05. A pattern speed of about 11 km s^{-1} kpc^{-1} is expected and the comparison of this with observation, for our Galaxy and for similar analysis related to other galaxies, is one of the tests of the theory. Comparisons between observed and predicted parameters, such as shape and velocity, do show reasonably good agreement (Lin, Yuan and Shu, 1969; Shu, Stacknik and Yost, 1971).

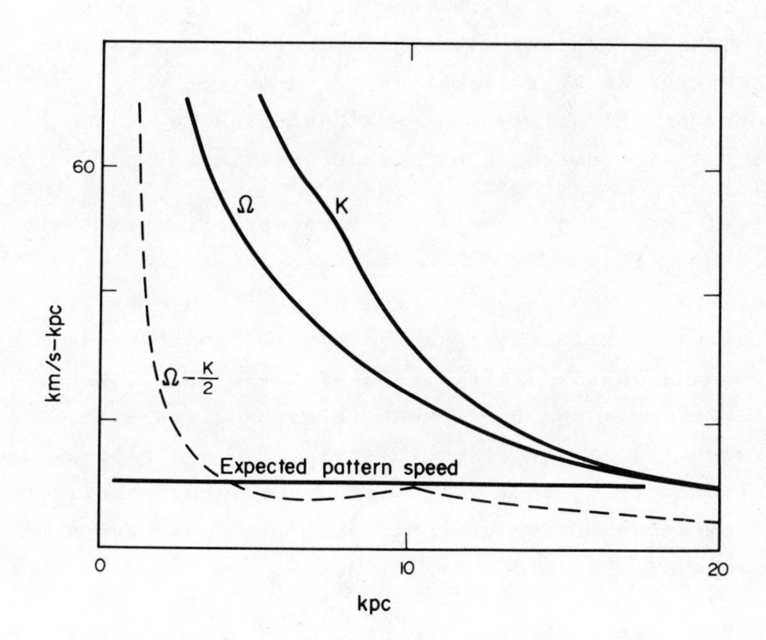

Fig. 9.05
$(\Omega - \kappa/_2)$ for our Galaxy showing an expected pattern speed of about 11 km s^{-1} kpc^{-1}.

The effect of a velocity dispersion can be seen by referring again to Fig. 9.01b. Suppose that the mass points are moving from left to right in that Figure not with a common velocity, as has been supposed, but with slightly different velocities. The bunching of masses coming out of the potential well will then generally be less pronounced and it will be superimposed on a 'random' bunching resulting from the non-uniform velocities. The density wave will thus become less effective as the velocity dispersion is increased.

It may be noted that resonances occur when κ/Ω is rational; in particular, when $\kappa = \pm 2\Omega$, equation (9.13) is zero and a standing density wave results, forming a ring instead of a spiral. This occurs at the "Lindblad resonances" which in our Galaxy are at about 4 kpc and 14 kpc from the centre. It is significant that the galactic spiral arms do not appear to extend further in than the smaller distance, which they would not be expected to do if they are due to density waves, and that the HII regions, which are indicative of star formation, extend from 4 to 13 kpc from the centre of the Galaxy. Note that these resonances correspond to $\nu = \pm 1$; co-rotation corresponds to $\nu = 0$.

The spiral arms do not contain a large fraction of the mass of a galaxy; Lin et al, (loc. cit.) estimate that the spiral gravitational field is only about 5% of the symmetrical field. The relative contributions to the spiral field by gas and by stars depends on the distribution of their velocity dispersions. Generally the gas has the lowest velocity dispersion, and if a relatively large mass exists in stars with a similarly small velocity dispersion they will contribute substantially to the spiral component of the gravitational field and hence to the maintenance of the spiral density wave.

An important matter that arises in the study of density waves is the dispersion equation - the relationship between the frequency ν and the wavelength λ, the latter giving the spacing of the spiral arms. In the early studies by Lin and Shu, λ was assumed to be small - that is, the arms tightly wrapped - so that powers of λ could be neglected.

The spacing of the arms may be expected to depend on several factors: the relative frequency of the spiral density waves, ν; the velocity dispersion v^2 in the material through which the wave passes; and its surface mass density σ. So we may write

$$\lambda = F\,(G, \nu, v, \sigma). \qquad (9.14)$$

where G is the constant of gravitation.

Dimensional analysis thus leads us to the reasonable conclusion that

$$\lambda\nu = \frac{v^2}{G\sigma}f. \qquad (9.15)$$

when f is a numerical factor which may be expected to be a function of the other parameters.

The formulation and solution of the function f has been pursued by Lin (1967), by Toomre (1969), and by Kalnjas (1970). The relation between wavelength and frequency found by Lin is illustrated in Fig. 9.06.

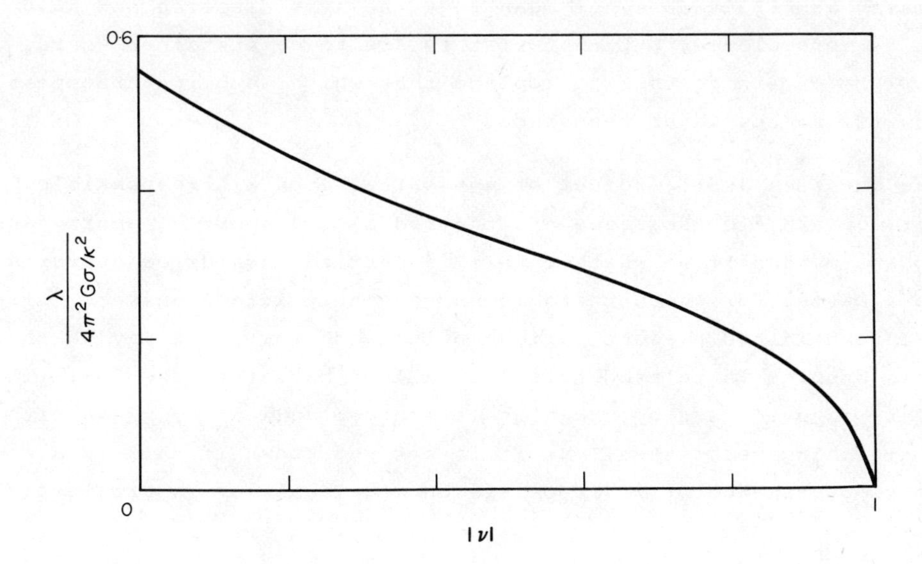

Fig. 9.06
Relation between frequency (in units of the epicyclic frequency) and wavelength λ (spacing) of spiral density waves according to Lin.

Kalnjas examines more open spirals (larger λ) with larger velocity dispersions, and finds much higher pattern speeds $\Omega_s \simeq 30$ km s^{-1}/kpc.

Two other points of interest arise from studies by Toomre (1969). He shows that Ω_s is not strictly constant; all trailing spirals tend to tighten and all leading ones to loosen. The effect is small but significant over 10^{10} yr. The density waves move inwards with a speed of about 10 kms^{-1} when they are wrapped as tightly as in our Galaxy; that is to say, they travel the radius of the spiral arm distribution - from 14 kpc into 4 kpc, in 10^9 yr.

Therefore in order to persist for 10^{10} yr, they must be constantly or repeatedly renewed. This refers to the group velocity of the waves, carrying energy; they are not of course "mass" arms; it does not mean that the arms disperse and have to be redeveloped, but that if they are to be sustained energy must be fed in to them to replace that which is being transported inwards to the inner resonance.

The question as to whether or not spiral arms are responsible for, or are a consequence of, the radial transport of energy and momentum, and if so whether this is carried inwards or outwards, is a matter for which there is no generally agreed answer. The references cited present various points of view. It may be that this problem is related to the infall of matter to the Galactic plane as evidenced by the high latitude clouds of hydrogen with approaching velocities, but it is not yet known if this is a local phenomenon or if it occurs on a galactic or intergalactic scale.

9.02 The Origin of Spiral Density Waves

These analyses show how spiral density waves may be maintained, but they do not show how they originate. There appear to be at least two possible origins of the initial perturbation which generates the spiral arm; a disturbance due to an assymetrical mass concentration at the outer edge, or a disturbance due to mass assymetry in the central region. These will be considered in turn.

Towards the outer edge of a galaxy the ratio gas/stars increases. The gas component generally has a small velocity dispersion and is therefore less stable than the stars. In addition, the mass and the patterns co-rotate in the outer regions, as can be seen from Fig. 9.05. Consequently not only is a disturbance in this outer region more likely to result in a growing perturbation, it will interact for a long period of time with any density wave which it induces. The co-existence of material and density wave spiral arms is foreseen by Lin, because despite the strong arguments given above that the overall pattern must result from the density waves, there are many lesser features which may be due to differential shear of material arms - such as the disconnected short spiral arm segments which protrude from the main spiral arms and given them a 'feathery' structure. When observed with high resolution in the longitude-velocity diagram of HI 21 cm radiation, spiral arms appear to be irregular spurs and pieces of arms that cannot be followed continuously for more than a few degrees (see the remarks by Kerr, by Weaver and by Westerhout in the review of the "Spiral Workshop" by Simonson, 1970). There are also indications that segments of spiral arms have different ages (van den Bergh, 1964).

Some support for the argument that density waves are induced by material perturbations in the outer parts of galaxies is provided by the study made by Simkin (1970) of the effects of a gravitational disturbance in a uniformly rotating disc of gas. She finds that the disturbance produces two spiral density waves which are coherent over a large part of the disc.

A different view is taken by Toomre, who argues (1969) that although satellites to galaxies may produce sufficiently powerful perturbations to be the cause of spiral arms, and he cites the Magellanic Clouds with our own Galaxy, and M51 with its satellite, as examples, satellites to galaxies do not occur commonly enough to be the general cause of spiral arms, the origin of which he suggests is to be found in circularly assymetrical distortions of the mass distributions in the centres of galaxies. Barred spirals are the evident examples of such a possibility. Kalnjas (1970) also produces convincing evidence that the spiral density wave structure may be driven by a bar-like mass distribution rotating in the centre, the junction of the arms to the bar probably being at the co-rotation point. In this case the spiral structure would extend from $\nu = 0$ to $\nu = +1$ and the arms carry energy outwards rather than inwards as in Toomre's study.

9.03 Ejection from Galactic Nuclei

Arp (1969a, 1969b) draws attention to a class of galaxies which possess a number of features as follows. They all have an open arm structure, type Sc or later, many indeed having a pair of almost straight arms which turn inwards only close to the nucleus. The arms show many bright knots and HII regions and therefore appear to be essentially young structures. One or both the arms often has a more than usually bright knot or a satellite galaxy at the outer extremity. Arp argues that these observations suggest that the 'satellites' at the ends of the arms have been ejected from the galactic nuclei and the arms are produced by matter trailing behind them which is wrapped by differential rotation into spirals. As the arms grow older they become more tightly wound - in accord with the indications of the photographs. The satellites have relatively short life-times, say a few 10^8 yr; before they disperse. Arp points out that these arms and satellites are isolated features, not more dense condensations in an otherwise uniform disc, and consequently appear to be material arms rather than density waves.

According to Arp, repeated ejections from galactic nuclei may occur at intervals $\sim 10^9$ yr, renewing the spiral arms.

His photographs in the papers cited provide convincing support for these arguments at least so far as a proportion of galaxies are concerned.

It is possible that somewhat similar effects may occur due to interaction between formerly independent galaxies approaching each other closely; if this is so the effect may be expected to occur more frequently in clusters of galaxies than in the general field.

9.04 Mass Arms

Although arguments were advanced in the introduction to this chapter demonstrating that many or most spiral arms could not be co-moving material structures and must be density concentrations moving through the material, the evidence gathered together by Arp and referred to in the last section also presents convincing evidence that some spiral arms are nevertheless material arms. It may well be that both types of arms are common; it is possible that material arms originate as Arp suggests and are the perturbations which lead to the development of density wave arms; it is thus possible that both types of arms co-exist in many galaxies.

Consequently it is worthwhile considering some of the earlier investigations into the properties of material arms, in particular the problems of the 'winding up' of the arms, and their stability.

9.041 'Winding-up'

As an example, consider the Andromeda galaxy M31. Measurements of the velocity of rotation of the HI gas at various distances from the centre have been given by Gottesman and Davies (1970). They are closely represented by a velocity increasing linearly with distance from the centre at the rate 5.6 km s^{-1} (arc min)$^{-1}$ to a value 230 km s^{-1} at 41 arc min, thereafter remaining constant at that velocity to the maximum distance from the centre reached, 150 arc min. The difference of the actual velocity from solid body rotation is then

$$\Delta V = 5.6\ \Theta - 230 \text{ km s}^{-1} \qquad (9.16)$$

where Θ is in arc minutes, and this differential velocity will cause any material arms to wind up. After time t the number of spiral turns about the centre made by a point at radial distance r is

$$n = \frac{\Delta V . t}{2\pi r} \tag{9.17}$$

Taking the distance to M31 to be 690 kpc, then 1 arc min = 200 pc. The number of turns on the spiral calculated by equations (9.17) and (9.16) is compared with observations in Table 9.01.

TABLE 9.01

Angular distance from centre of M31 arc min	n predicted 10^{10} yr.	n predicted 10^{9} yr.	n observed
0	0	0	0
10	0	0	1
30	0	0	1½
50	8	1	2
70	18	2	2½
90	24	2½	3
110	28	3	
130	30.5	3	
150	32	3	

It is apparent that agreement is fairly good only if the age of the arms is ~ 10^9 yr, about 10% of the probable age of that galaxy.

Since a considerable proportion of large galaxies have spiral arms, typically with only a few turns as in the case of M31, it follows that for the spiral structure to be caused by differential velocity shear of matter arms, the arms must be recreated at intervals of about 10^9 yr. If the arms are produced by the ejection of material from galactic nuclei as argued by Arp, this also indicates the time interval between ejections.

9.042 Stability

If the arms are bars of matter wound into spiral shapes by differential rotation, then the question arises as to whether or not the matter is expected to disperse and if so how the time scale of dispersal compares with the 10^9 yr lifetime of the arms derived in the previous section.

In our Galaxy, the arms are about 800 pc wide and spaced 2 kpc apart in the solar region. Cool gas clouds have a velocity dispersion of $\pm$ 5 km s^{-1} or a little greater. Consequently the arms (represented by clouds illuminated by stars recently formed in them) will disperse in ~ 2.10^8 yr, which is the time for the clouds to travel 1 kpc, unless the arms are bound together by some force. Table 9.01 shows that in M31, for example, the spiral arms would only have achieved half a turn in that time, one sixth of the winding observed, and only the minority of galaxies with a very open arm structure such as shown on some of Arp's photographs could have arms so young.

Although gravitational forces such as are effective in gravitational waves affect both stars and gas clouds equally, pressure forces act selectively on the gas only, on account of the very much smaller collision cross section of stars in proportion to their masses. However, as has been pointed out earlier, spiral arms appear to be essentially formations of clouds, those stars that have a similar distribution being there only because they have recently (within 10^7 yr) formed in the clouds. On the other hand, the larger velocity dispersions of older stars reduces the effect of spiral density waves on their distributions to insignificant levels. Consequently the selective concentrations of clouds and young stars into spiral arms does not help us to distinguish between pressure forces or spiral density waves as the maintaining element involved.

The size of a pressure force required to prevent dispersion of matter arms can be estimated as follows. The pressure due to the random velocities tending to disperse the material is

$$P_v = \frac{1}{3} \rho v^2 \tag{9.18}$$

where ρ is the gas density in the arms and v is the r.m.s. velocity of the gas clouds. From chapter 7 typically $\rho = 3 \times 10^{-24}$ g cm^{-3} and $v \simeq 7 \pm 2$ km s^{-1}. Hence

$$P_v \approx 5 \times 10^{-13} \text{ dyn cm}^{-2} \tag{9.19}$$

If the material in the spiral arms is held together by the pressure of an interarm plasma, then for pressure balance the plasma pressure must have a value similar to equation (9.19).

The temperature and density of the interarm gas are estimated as ~ 1,000°K and 0.3 H atoms cm^{-3} (chapters 7 and 8), from which

$$P = nkT = 4 \times 10^{-14} \text{ dyn cm}^{-2}, \tag{9.20}$$

an order of magnitude too low. However, in order to be effective in preventing dispersal of the gas clouds making up a spiral arm, interarm pressure must be prevented from penetrating between the clouds and the most likely barrier is a magnetic field. To be effective this would have to exert a pressure

$$P = \frac{H^2}{8\pi} \sim 5 \times 10^{-13} \text{ dyn cm}^{-2};$$

$$H \sim 3.5 \times 10^{-6} \text{ gauss} \tag{9.21}$$

which is similar to the values currently estimated from interstellar Faraday rotation, dispersion, and Zeeman splitting measurements (chapter 7).

In addition to the strength of the field having to be of this amount a systematic configuration is necessary and the most stable configuration is probably that of a helix wound around the spiral arms (Hoyle and Ireland, 1961; Ireland, 1961). Such a field may be regarded as having two components, one along the spiral arm, and the other wrapped circularly around the arm perpendicularly to its axis. The former resists transverse shearing of the arm - binds it together along its length; the latter prevents dispersal of the arm material and resists shearing along the arm due to differential rotation of the galaxy. Hoyle and Ireland show that the pressure exerted by a magnetic field of 5×10^{-6} gauss is able to withstand this shearing.

9.05 Spiral Arms in Relation to Star Formation

It was noted in the first half of this book that stars form in the more dense interstellar clouds and that these clouds are gathered into spiral arm formations. The question arises as to whether or not the cloud densities are increased and star formation thereby assisted, by the action of forces producing or resulting from spiral arm formations. Particularly one may think of compression forces such as those maintaining the stability of matter arms, or shock waves produced by the movement of density waves through the galactic disc of gas.

In the case of compressive forces, the differences in density between arm and interarm regions may be very high in the pressure stability situation discussed in chapter 8, the density ratio being perhaps $10^3 : 1$. Shock waves appear likely to produce relatively small increases in density, by factors varying up to ten (Roberts, 1969) and the energy dissipated in producing them would cause the density wave shock to die out in 10^9 yr - time for three rotations of the galaxy (Kalnjas, 1970, 1972). However, as Roberts points out, the additional compression may be effective in causing massive clouds to become unstable against contraction under self-gravitation, even though it would not directly cause fragmentation and compression of the fragments.

Therefore he argues that the shock may be instrumental in triggering star formation by causing the initial cloud contraction. Newly formed stars would then be along the inner edges of the arms, a situation for which there may be some observational support. Matthewson et al, (1971) have made a 1415 MHZ survey of the galaxy M51 which shows strong radio "arms" coincident with dust lanes along the inner edges of bright arms, indicating compression of the gas and the magnetic field locked in it. The displacement of the dark radio edges from the bright arms would give a time for star formation of 6×10^6 yr using the wave speeds calculated as in section 9.01. This is the free fall collapse time for a gas cloud beginning with a density $n(H) = 25\ cm^{-3}$, a reasonable value. Similarly, photographs of M33 in different colours by Courtes and Dubout-Crillon (1971) show an age progression from young stars with HII regions along the inner edges of the arms to stars about 8×10^6 yr old at the outer edge (expanded OB associations, for example). This indicates a density wave with $\Omega_p = 15\ km\ s^{-1}$. Further support is given by 21 cm H line observations of M31 (Guibert, 1974) which show that the angular velocity of the spiral pattern at 10 kpc from the centre is 11 km s^{-1} kpc^{-1}, equal to the circular velocity of the outermost HII regions.

Chapter 10.

CONDITIONS FOR CONTRACTION OF CLOUDS.

It is pertinent to begin this chapter by quoting from two paragraphs in the introduction:

> "The process of forming a star may be said to have ended when a mass of gas has reached a state in which, excepting unusual accidents, it will inevitably become a star of essentially that same mass. Thereafter, it becomes the concern of stellar evolution rather than of star formation.
>
> "It is more difficult to define a starting point for the formation of a star. The formation of galaxies not being a part of our present study, it is reasonable to use as a beginning the existence of a cloud of interstellar matter of galactic mass ....."

This interstellar matter is observed to have a cloud-like structure, and the more dense clouds are gathered into spiral arms; the reasons for this have been considered in chapters 8 and 9. It was shown in chapter 2 that the spiral arms are the location of star formation and the spatial relationship between different kinds of clouds and stars provides strong circumstantial evidence that stars form in the colder, more dense, quiescent clouds in which the gas is probably largely molecular. Having regard to the first paragraph quoted above, it is necessary to determine the conditions under which a cloud will contract until it is set inevitably on the route to becoming a star.

For a cloud to contract, the internal expansion forces - the sum of the various pressures and centrifugal force - must be less than the contractive forces - the sum of external pressures and self gravitation. The Virial Theorem is a statement of these forms. Consider an assembly of particles - atoms, or molecules, or stars, etc. - each particle having mass m and position vector $\overline{r}$, and being subject to a force $\overline{F}$. Then acceleration of the moment of inertia of the assembly of particles is represented by

$$\frac{d^2 I}{dt^2} = \frac{d^2}{dt^2} \sum (m\overline{r}^2)$$

$$= \sum 2 \, (m\dot{\overline{r}}^2 + \overline{r}. \, \overline{F}) \qquad (10.01)$$

That is,

$$\frac{1}{2} \, \frac{d^2 I}{dt^2} = 2(\text{Kinetic energy}) + \sum \overline{r}. \, \overline{F} \qquad (10.02)$$

The term $\sum \overline{r}. \, \overline{F}$, which is a measure of the forces of interaction between the particles, is Clausius' Virial (Clausius, 1870) and was derived by him in this form in an attempt to account for forces of interaction between gaseous molecules as was done differently by van der Waals.

The Virial is analogous to the work done by the forces $\overline{F}$ in moving the particles to $\overline{r}$ and therefore represents the potential energy in the system; so we can write

$$\frac{1}{2} \, \frac{d^2 I}{dt^2} = 2(\text{Kinetic energy}) + \text{"Potential energy"} \qquad (10.03)$$

If the assembly of particles is in equilibrium - if there are no net forces acting to change the distribution of the assembly of particles as a whole - then there will be no acceleration of the moment of inertia of the assembly, that is, d^2I/dt^2 will be zero. Particles may be redistributed locally within the assembly without disturbing the overall equilibrium. On the other hand, if net forces are acting to expand the assembly, then the moment of inertia will be accelerated towards larger values, $d^2I/dt^2 > 0$; whereas forces causing contraction are represented by $d^2I/dt^2 < 0$. So we may write:

$$2(\text{Kinetic energy}) + \text{Potential energy} \begin{Bmatrix} > 0; \text{ expansion} \\ = 0; \text{ equilibrium} \\ < 0; \text{ contraction} \end{Bmatrix} \quad (10.04)$$

There are several sources of energy within interstellar clouds, and various forces acting upon them, and these will now be considered with a view to applying the criterion (10.04), in particular to determine the conditions necessary for clouds to contract to form stars.

Sources of kinetic energy are the thermal motions of the cloud particles, rotation and turbulence; sources of potential energy are gravitation, radiation and magnetic fields.

If the assembly of particles is a cloud of perfect gas with temperature T, the kinetic energy per particle is $\frac{3}{2}$ kT and the total thermal kinetic energy K is

$$K = \frac{3}{2} NkT = \frac{3}{2} PV \quad (10.05)$$

Assuming a spherical and homogeneous cloud of density ρ, mass M and radius R rotating with an angular speed ω, the kinetic energy of rotation is:

$$K_\omega = \frac{1}{2} I\omega^2 = \frac{1}{5} MR^2\omega^2 \quad (10.06)$$

The average kinetic energy of turbulence per unit volume is $\frac{1}{2} \rho \overline{v_t^2}$ where v_t is the r.m.s. turbulent velocity. The total kinetic energy of turbulence for the whole cloud is

$$K_t = \frac{1}{2} M\overline{v_t^2} \quad (10.07)$$

Turning to the sources of potential energy, the gravitational potential energy Ω of a homogenous sphere is

$$\Omega = -\frac{3}{5}GM^2/R \qquad (10.08)$$

The energy density of the radiation field is aT^4 where a is the radiation density constant, and so the total radiant energy in the cloud Ω_r is

$$\Omega_r = \frac{4}{3}\pi R^3 aT^4 \qquad (10.09)$$

The magnetic energy density is $H^2/8\pi$ where H is the r.m.s. magnetic field strength, and the total magnetic energy in the cloud Ω_m is thus

$$\Omega_m = \frac{1}{6}R^3H^2 \qquad (10.10)$$

If there is an external pressure P_s acting on the surface of the cloud, due to intercloud gas, radiation and magnetic fields, this may be allowed for by using the net pressure in equation (10.05) and writing

$$K = \frac{3}{2}(P - P_s)V \qquad (10.11)$$

Condition (10.04) now becomes

$$2(K + K_\omega + K_t) + \Omega + \Omega_r + \Omega_m \begin{cases} < 0; \text{ contraction} \\ = 0; \text{ equilibrium} \\ > 0; \text{ expansion} \end{cases} \qquad (10.12)$$

Substituting equations (10.06) to (10.11) and rearranging, this gives as the condition for equilibrium or contraction

$$4\pi\rho\left\{\frac{k}{m}T + \frac{1}{3}\left(\frac{2}{5}(R_\omega)^2 + \overline{v_t^2}\right)\right\} + \frac{4}{3}\pi aT^4 + \frac{1}{6}H^2 \leqslant \frac{3}{5}G\left(\frac{M}{R^2}\right)^2 + 4\pi P_s \qquad (10.13)$$

where m is the average mass of a cloud particle.

This can now be applied to interstellar clouds, for which values of temperature, density, size, magnetic field and motions are gathered together in Part 1. In particular, we are concerned with those cold, dark clouds which appear to be the places where stars form. It is informative to consider the terms in (10.13) separately, as follows.

(a) Gas pressure, turbulence and rotation Estimates of the motions in dark clouds (chapter 7) made from doppler shifts of spectral lines at both optical and radio wavelengths, do not distinguish turbulent motions from differential rotation within the cloud. It is to be expected, however, that the sizes of these motions will be similar and therefore it is reasonable to write $v_t^2 \approx \frac{1}{2}v_{obs}^2 \approx (R\omega)^2$ so that

$$\frac{1}{3}\left(\frac{2}{5}(R\omega)^2 + \overline{v_t^2}\right) \approx \frac{1}{4}\overline{v}_{obs}^2 \qquad (10.14)$$

The observed velocities reported in chapter 7 are $\leqslant$ (0.5 to 0.75 km s^{-1}); hence

$$\frac{1}{3}\left(\frac{2}{5}(R\omega)^2 + \overline{v_t^2}\right) \approx 10^9 \text{ erg} \qquad (10.15)$$

The corresponding kinetic temperatures in the clouds deduced from the excitation of radio wavelength spectrum lines of atoms and molecules, were found to be about 5K and consequently

$$\frac{R}{\mu}T \approx 4 \times 10^8 \text{ erg} \qquad (10.16)$$

Since turbulence decays to thermal energy which is radiated away, it would be reasonable to expect that equation (10.15) would be comparable with equation (10.16); this would be so if the velocities in the cold clouds are $\approx$ 0.4 km s^{-1}, which is consistent with the observations.

(b) Radiation pressure The effective temperature of the cosmic background radiation is $\approx$ 3K, and hence

$$\frac{4}{3}\pi \, aT^4 = 2.4 \times 10^{-12}$$

However, a corresponding pressure is included in P_s and cancels.

(c) Magnetic pressure The estimates of the strengths of interstellar magnetic fields, reported in chapter 7, may be summarised as follows:

Region	n_H cm^{-3}	H gauss
interarm	0.1	10^{-6}
cloudlets	1-2	10^{-5}
dense HI clouds	$10^{3\pm1}$	10^{-5} - 4×10^{-5}
dense molecular clouds	$\geqslant 10^6$	6

These fields may also act as strongly in P_s as in P (see equation (10.11)) if the ambient density around the cloud is high, and may indeed become ineffective within the cloud if the ionisation level falls so low that the material decouples from the cloud. Ionisation by a large flux of low energy cosmic rays would prevent this happening quickly, but if the interstellar gas is heated by the less penetrating x-rays and γ rays (chapter 8) the interiors of dense clouds will have a very low electron density and will therefore decouple (Nakano and Tademaru, 1972). In the present state of the data perhaps the best that can be assumed is that the magnetic fields inside and immediately outside proto-fragments of clouds are not very different prior to fragmentation, but the magnetic field inside the cloud as a whole may be higher, by a factor up to four, than in the surrounding gas in the spiral arm. Therefore since we are concerned mainly with fragmentation and the contraction of fragments, the magnetic pressure also contributes to P_s as well as to P.

Having regard to (a), (b) and (c), condition (10.13) probably reduces in practice to:

$$4\pi \; 10^9 \frac{\rho}{\mu} \leqslant \frac{3}{5} \; G \; (\frac{M}{R^2})^2 \qquad (10.17)$$

which is

$$M \geqslant 3.5 \times 10^{-9} \; \mu^{-3/2} \; \rho^{-\frac{1}{2}} \; M_\odot \qquad (10.18)$$

Otherwise the minimum density of a mass M for contraction is

$$\rho \geqslant 1.3 \times 10^{-18} \; \mu^{-3/2} \; (\frac{M}{M_\odot})^{-2} \; g \; cm^{-3} \qquad (10.19)$$

The lower limit to the masses of stars formed appears to be 0.03 $M_\odot$, the average stellar mass 0.12 $M_\odot$ (chapter 3). For cloud fragments with these masses to be able to contract, by condition (10.18) they must be formed with densities $\geqslant 5 \times 10^{-15}$ g cm^{-3} and $\geqslant 3 \times 10^{-16}$ g cm^{-3} respectively, assuming the gas to be molecular hydrogen; the corresponding number densities of molecules are 1.7×10^9 cm^{-3} and 10^8 cm^{-3}. Referring to Fig. 7.04, such densities are found in the ambient nebulae around OH sources and in infrared nebulae, all of which are considered to be protostars or to be associated with them. They are therefore reasonable densities.

The largest masses that can form stable main sequence stars, limited by radiation pressure, are expected to be 60 $M_\odot$; these could contract at a density $\rho \geqslant 1.3 \times 10^{-22}$ g cm^{-3} corresponding to $n(H_2) \geqslant 40$ cm^{-3}. These and higher densities are found in the clouds but not in the cloudlets in spiral arms, or in the interarm regions (Fig. 7.04).

The conditions for contraction of protostar clouds are therefore consistent with the observations, chapter 2, that star formation is confined to the dense, dark clouds in spiral arms. Furthermore, since at these densities much of the hydrogen will combine to form molecules when the temperatures of the interstellar grains on which they form are $\leqslant$ 20 K, (Lee, 1972) the association of abundant molecules with regions of star formation is also to be expected.

The contraction of an interstellar cloud as a whole to high densities does not proceed homologously; the central ten per cent or so of the mass contracts rapidly and the outer parts more slowly, so that very great differences in density between the inner and outer parts result (Kikuchi, 1967). This may produce a situation where the condition for contraction (10.13) allows only massive stars to be produced outside the central core in which dwarf stars can also form. This would fit the picture of groups of T-Tauri dwarf protostars embedded in larger aggregates of massive OB stars, which the observational data appears to present (chapter 2).

Doubts have been expressed as to whether the high densities required for contraction of stellar masses can be reached except in massive clouds, which would imply that stars can only form in large aggregates. However, some young stars appear to have formed in isolation, and to account for their contraction at the lower densities expected in isolated clouds of such low mass, Hunter (1972) has suggested that inwardly directed velocities in the internal cloud motions may be responsible. Referring to equation (10.13), we have assumed that velocities of internal turbulence are randomly directed and consequently tend to disperse the gas, effectively producing an internal pressure leading to expansion. In any finite situation however, in which the number of turbulent cells is not very large, a situation will sometimes occur in which most of the cells will be moving inwards; the term $+\,\overline{v_t^2}$ is then replaced by a negative term $-\left[\overline{v_t \text{ (net inwards)}}\right]^2$. If these net inward velocities are as high as 0.7 km s^{-1}, then the term

$$\left\{\frac{k}{m} T + \frac{1}{3}\left[\frac{2}{5}(R\omega)^2 - \left[v_t \text{ (net inwards)}\right]^2\right]\right\}$$

becomes zero and there is no lower limit to the mass of a cloud which can collapse at any density. However, the energy of these inwardly directed velocities compresses the material and heats it, and this heat must be radiated away quickly enough to prevent the temperature and hence the gas pressure rising substantially.

Although this hypothesis has attractions in possibly accounting for the formation of isolated stars, it fails to account for the frequency distribution of stellar masses and could therefore apply only to a very small number of cases. That is to say, it cannot be the mechanism by which stars in general are formed. The condition (10.13), applied to the properties of interstellar clouds as determined in chapter 7, shows that protostars of all observed masses can form in the more dense clouds occurring in spiral arms, and observations of the locations of star formation described in chapter 2 indicate that this is indeed where stars form.

Chapter 11.

THE FRAGMENTATION OF CLOUDS INTO PROTOSTARS AND THE FREQUENCY DISTRIBUTION OF PROTOSTAR MASSES.

11.01 Introduction

The manner in which an interstellar cloud breaks up to form fragments which subsequently contract to become stars is the crux of the problem of star formation. It is to be expected that it determines the frequency distribution of stellar masses and the rate of star formation, and the observational data relating to these matters must be accounted for by any theory of the process of star formation if that theory is to be accepted. There is probably nothing to be gained by considering theories which do not attempt to meet this criterion and therefore such theories are not dealt with in any detail in this chapter, although some of them will be referred to.

The various theories which seek to account for the origin of protostars can perhaps be divided among three types; those which suppose that protostars are relics of an earlier expansion phase in the Cosmos, being objects that were sufficiently dense to overcome the expansion locally; those that regard protostars as fragments of an explosion or at least a direct consequence of the compression produced by an explosion; and the majority in which protostars are condensations of increasing density in less dense clouds.

Layzer (1964) supposes that clusters and galaxies of protostars are the result of a hierarchy of clustering in the early life of the Universe, volumes of gas expanding more slowly than the average in the Cosmos. The protostars which still remain and lead to continuing star formation have been inhibited from contracting during the 10^{10} yr since they were formed, but at intervals some become unstable through the effect of external pressure (see condition (10.13)) produced by an HII region occurring in their vicinity.

The explosive fragmentation of a massive condensation has been examined by Bird (1964; 1968). By applying the Virial Theorem (see condition (10.13)) to an interstellar cloud of mass $10^{5\pm1}$ $M_{\odot}$ in which the central part is colder due to shielding by the opacity of the outer parts, he shows that a central condensation of 10^2 to 10^3 $M_{\odot}$ forms and contracts to stellar densities. However, it is pulsationally unstable and explodes. The ejected fragments of the massive object and its parent cloud are supposed to be protostars of expanding associations of OB stars.

Another explosive origin for stars is suggested by Abartsumyan (1964), based on the proposition that in addition to the diffuse matter in the galaxies there exists super dense matter, with a density comparable with that of the nuclei of atoms, which can change state into ordinary matter. This change occurs in an explosive manner and leads to expanding associations of young stars, and is the cause of subsequent energetic outbursts by young stars.

A hierarchy of turbulent eddies, covering the whole spectrum of dimensions from clusters of galaxies through galaxies and stars to viscosity in the interstellar gas, was proposed by von Weizsacker (1951). The turbulent energy decays to heat through the hierarchy, to be radiated away. The time scale for the decay of the initial turbulence in a galaxy is, however, $t(\text{decay}) \approx l/v$ where l is the average dimension of a turbulent eddy and v the turbulent velocity; an upper limit is given by the reciprocal of the typical shearing velocity gradient due to galactic rotation, $\approx 5.10^7$ yr. Therefore only stars formed in the earliest period of evolution of the galaxy could do so as a consequence of the initial turbulence; but turbulence will continue if dynamical energy is fed into the gas on a large scale and it is significant, as noted in chapter 10, that the turbulent energy in the interstellar gas is still of the same order as the thermal energy. However, as noted in that chapter, turbulence does not generally assist compression, and there is no observational evidence to support the supposition that stars form in regions of high turbulence; on the contrary, young stars appear to form in the most quiescent clouds (chapters 2 and 7), and when there is violent turbulence in the interstellar gas, for example in young HII regions, it appears to be a consequence of star formation rather than a cause of it, because young stars are concentrated to the centre of such regions and star formation does not spread out from the centre with the turbulence.

Those theories which rest on the basis that protostars result from condensation and fragmentation of interstellar clouds can themselves be divided into different types according to the process by which fragmentation is supposed to occur, for examples, condensation due to statistically random density fluctuations; a hierarchy of successive fragmentations into smaller masses at increasing densities; and fragmentation due to the physical effects of the condensation of atoms to form molecules or solids in the clouds. All of these which lead to predictions of the properties of the protostars formed, including the frequency distribution of masses for comparison with the observed data, will be dealt with in some detail in turn.

11.02 Random Fragmentation

The possibility that massive clouds may break up into fragments in a statistically random manner has been examined by several authors: Kushwaha and Kothari (1960), Auluck and Kothari (1965), Kruszewski (1961), and Kiang (1966) suppose that the volumes divide randomly; Marcus (1968) that the masses split randomly; while Larson (1973) assumes a random probability of fission of condensations into two subcondensations occurring repeatedly in a hierarchical manner as overall cloud contraction proceeds. Although there is no substantial physical basis for any of these hypotheses, they are worth examining because agreement between predicted and observed mass functions may indicate that the process has some basis in fact.

11.021 Volume Fragmentation

The study by Auluck and Kothari (1965) begins by considering the random fragmentation of a line of length ℓ into N parts; the average number of fragments with length $\geqslant$ x is

$$N(x) = N(1 - x/\ell)^{N-1} \qquad (11.01)$$

This is now applied to the division of a rectangular parallelopiped by planes parallel to the sides intersecting the edges into random lengths. This leads to an expression for the number of fragments n (V) with volumes in unit range about V,

$$n(V) = \frac{4N_O}{V_O} \int_0^\infty \frac{K_O(y)}{y} \exp\,(-4a/y^2)\,dy \qquad (11.02)$$

here N_O is the total number of fragments and V_O the average volume of a fragment (so that $N_O V_O$ is the total volume of the cloud), $a = V/V_O$, and $K_O(y)$ is the Bessel function of imaginary argument. Assymptotic solutions are found for $a > 1$ and $a \to 0$ and, assuming a uniform mass density in the cloud, the function (11.02) represents the number of stars formed per unit range in mass, F(M). Over a range of 100:1 in mass, the values of F(M) predicted by this theory change by a factor 4×10^7; the observed change (Fig. 11.01) is by a factor 5×10^4, and the theory therefore does not account satisfactorily for the observations.

Kruszewski (1961) reconsidered the volume fragmentation theory following the work of Kushwaha and Kothari, applying as a limitation Jean's criterion - that for gravitational contraction the dimension of a density perturbation must be larger than

$$\lambda_O = v/\sqrt{4\pi G\rho} \qquad (11.03)$$

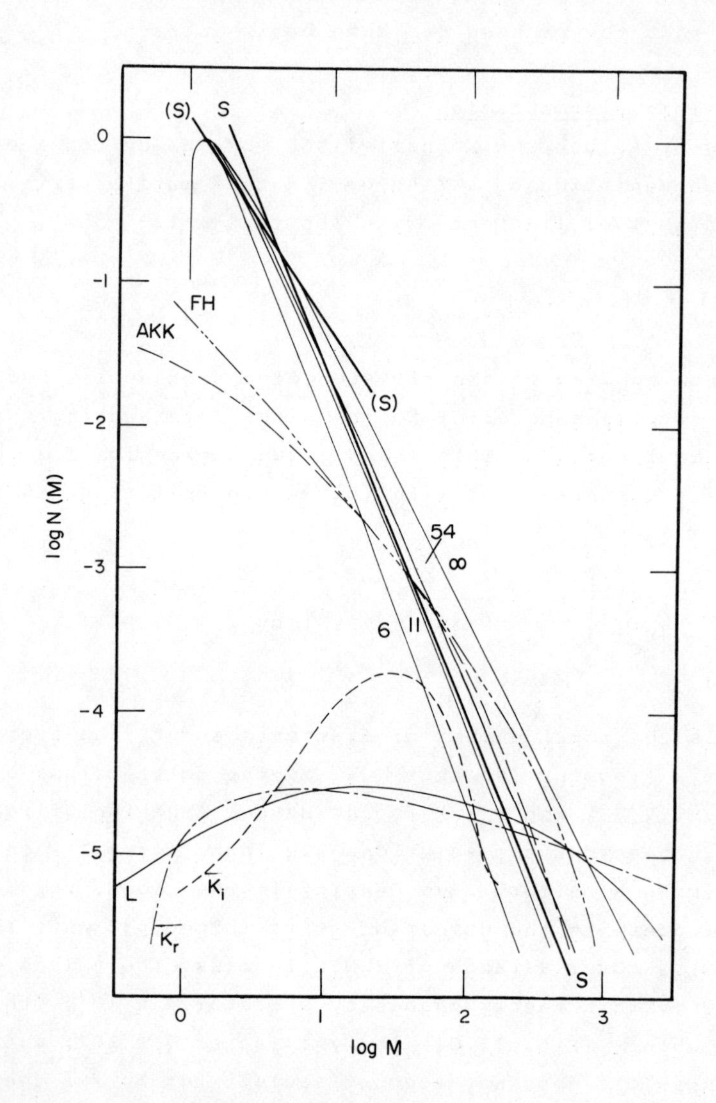

Fig. 11.01
The frequency distributions of the masses of stars formed, predicted by various theories, compared with observational data. N(M) is the relative number of stars formed in unit range of mass at mass M. The theories described in the text are due to Fowler and Hoyle (FH), Auluck, Kushwaha and Kothari (AKK), Kiang (Ki), Kruszewski (Kr), Larson (L), Reddish (curves marked 6, 11, 54 and infinity, see text). The observational data (S) are due initially to Salpeter (and see Fig. 3.04).

where v is the velocity of sound and ρ the density, from which it follows that the minimum mass that can contract is given by

$$M \geqslant \rho\lambda_0^3 = 6.6 \times 10^{-10}\ v^3\rho^{-\frac{1}{2}} \qquad (11.04)$$

It is assumed that the probability distribution of density perturbations does not depend on the linear extent of the perturbation, and hence the probability of a perturbation of extent λ is inversely proportional to the volume of the perturbation, λ^{-3}. Thus

$$P(\lambda)\ d\lambda = 2\lambda_0^2\ \lambda^{-3}\ d\lambda \qquad \lambda_0 \leqslant \lambda \leqslant \infty \qquad (11.05)$$

Applying this probability distribution to three independent dimensions with the criterion (11.04) leads to the predicted frequency distribution of protostar masses

$$F(M) = \tfrac{1}{2}\ (\ln M)^2\ e^{-(\ln M)}\ d(\ln M) \qquad (11.06)$$

where M is in units of $M_0 = \rho\lambda_0^3$.

This gives better agreement with the observed data than do the theories by Kushwaha and Kothari (loc. cit.) and Auluck and Kothari (loc. cit.) although still showing discrepancies by an order of magnitude for stars with masses below about 0.1 suns.

Kiang (1966) improves upon earlier theories of volume fragmentation in which the volumes were produced by sets of intersecting planes, which gave unrealistically common surfaces to many cells. Kiang's theory is based upon a random distribution of points, each point being surrounded by a cell which forms the volume fragment. The way in which the walls of the cells are defined can be understood easily by considering the process forming areas on a plane, Fig. 11.02. The boundaries are the perpendicular directions of lines joining a point to its neighbours, the junctions of bisections being equidistant from the three points forming a pair of lines.

The frequency distribution of the volumes of the cells was found approximately by calculating numerical examples, and indicated that the accurate solution is

$$F(V) = \frac{c}{\Gamma(c)} (cV)^{c-1} e^{-cx}; \quad c = 6 . \tag{11.07}$$

This distribution differs greatly from the initial stellar mass function (Fig. 11.01) and it is therefore concluded that it does not represent the course of events in star formation.

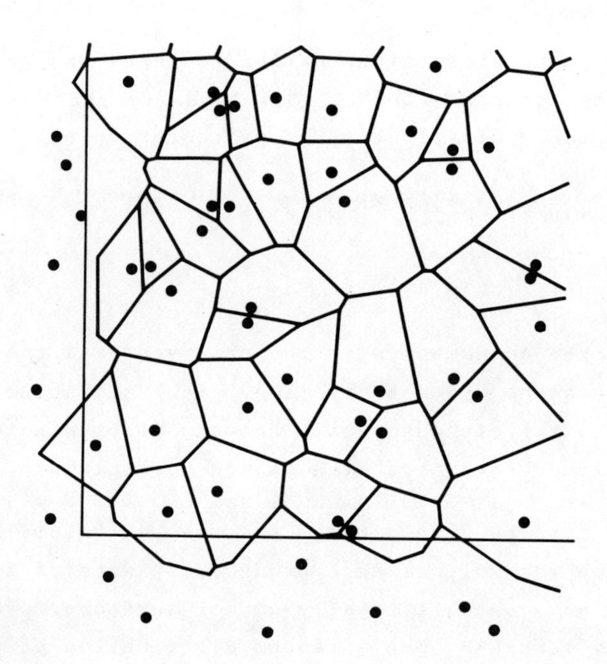

Fig. 11.02
The mode of fragmentation according to Kiang.

11.022 Mass Fragmentation

A general theory of particle splitting due to Filippov is applied by Marcus (1968) to the fragmentation of interstellar clouds. He also reviews the observational data and other theories.

The theory is based on the following assumptions.

The probability that a cloud of given mass will split into a number of smaller masses is a function (assumed to be a power function) only of the mass of the cloud and is proportional to the time interval. The number and masses of the fragments are random variables. It is assumed that large masses break up more readily than small ones, and that the fragments are distributed among all masses down to some lower limit so that not all the mass goes into ultimately small fragments or into fragments of any one mass. Furthermore it is assumed that the process is independent of the mass of the original cloud.

If the dependence on mass of the probability of splitting is

$$P(M) = M^n , \quad n > 0 ; \qquad (11.08)$$

if the fraction of the mass of the original cloud which goes into fragments with masses less than M is

$$f(M/M_0) = (M/M_0)^k , \quad k > 0 \qquad (11.09)$$

and if fragmentation begins at t = 0, it follows from the above assumptions that the number of fragments in the mass range M to M + dM at time t is

$$F(M) = \frac{\partial}{\partial M} \left\{ te^{-t} \int_0^M F(1 + \frac{k}{n}, 2, (1-(t+1)^{\frac{k}{n}} M^k)t) dv \right\} \qquad (11.10)$$

where $v = (t + 1)M^n$

and

$F(\alpha,\beta,\gamma)$ is the confluent hypergeometric function.

Marcus studies in detail the cases $k = n, 2n, 3n$, etc., at various t, and the mixing of mass spectra with different t to simulate the effect of a conglomeration of fragments from clouds breaking up at different epochs.

The results are presented in a number of Figures, which give

$$q(M) = \frac{1}{M} F(M) \tag{11.11}$$

The slopes $dq(M)/dM$ of the curves are well distributed over the range from about 0.75 to 1.9 as the parameters (k, n) take various values, and it is clearly possible to choose values for the parameters to give the observed slope 1.35 (chapter 3). However, only if the parametric values have some physical basis can Marcus' theory be justified and so far none has emerged: the validity of the theory therefore remains in doubt.

A theory having elements of both a random and a hierarchichal fragmentation was developed by Larson (1973). It is assumed that any cloud in an aggregate of clouds has a certain probability, independent of the previous history of the cloud, of fission into two equal masses. This probability is a free parameter which has to be adjusted to give agreement with observation. The result is closely approximated by a gaussian distribution for the fraction of mass in stars per unit logarithmic interval in stellar mass. Thus

$$F(M)\ M\ d(\log M) = F(M)\ dM \tag{11.12}$$

is gaussian. As can be seen from Fig. 11.01, this does not fit the observational data.

11.03 Hierarchichal Fragmentation

In a study of the processes of star formation, Fowler and Hoyle (1963) put forward the hypothesis that fragmentation of a cloud into stellar masses occurs in a series of steps, equal amounts of mass being left behind at each step. Thus

$$N_n M_n = \text{constant} \tag{11.13}$$

where N is the number of protostars of mass M formed in step n. If the density increases by a factor f^2 at each step, it follows from the condition for the condensation of fragments (10.19), which may be written

$$M \propto \rho^{-\frac{1}{2}} \tag{11.14}$$

that the mass of a fragment decreases by a factor f at each step. Hence if M_c denotes the mass of the cloud,

$$\frac{M_c}{M_n} = f^n \tag{11.15}$$

The total number of fragments with masses greater than M_n is

$$N(M > M_n) = \sum_{n=1}^{n} f^n = f^{n+1}/(f-1) \propto f^{n+1} \tag{11.16}$$

which using equation (11.15) becomes

$$N(M > M_n) \propto (M_c/M_n)^{\frac{n+1}{n}} \tag{11.17}$$

Since from equation (11.15)

$$n = \ln(M_c/M_n)/\ln f \tag{11.18}$$

it follows that the number of fragments with masses greater than M becomes

$$N(>M) = \text{constant } M^{-(1+\ln f/\ln (M_c/M))} \tag{11.19}$$

The numbers of fragments of various masses can then be calculated, giving for the number of protostars per unit range in mass

$$F(M) \propto M^{-\beta} \tag{11.20}$$

where

$$\beta = 2 + \ln f/\ln (M_c/M),$$

M_c being the mass of the cloud. The value of β varies from 2.4 at one solar mass to 2.7 at five; the observed index being close to 2.35. Outside this range of masses however, the agreement is not good and the formula underestimates the number of bright stars by nearly three orders of magnitude (Reddish, 1965; Fig. 11.01).

The possibility that masses are built up by coalescence following fragmentation has been examined by Nakano (1966). To make the problem tractable he presumes inelastic collisions (no re-fragmentation), takes account of binary collisions only, and assumes all masses involved to be integral multiples of some unit mass.

The number of fragments of mass M is increased by collisions between fragments of masses M^1 and $M-M^1$; all other collisions reduce the number. Thus,

$$\frac{\partial N(M,t)}{\partial t} = \frac{1}{2} \sum_{M^1 = 1}^{M-1} C(M^1, M-M^1) N(M^1) N(M-M^1)$$

$$- N(M) \sum_{M^1 = 1}^{\infty} C(M, M^1) N(M^1) \tag{11.21}$$

where

$$C(M,M^1) = \sigma(M,M^1)\ \overline{v}(M,M^1)$$

is the collision rate, σ being the effective collision cross section and $\overline{v}$ the mean relative velocity between the colliding fragments.

Equation (11.21) was integrated numerically for various initial distributions to give a frequency distribution of masses which was found to reach a limiting form after a time interval equal to a few collision times. The results showed that the final form of the mass function did not depend on the initial form at $t = 0$ but depended quite markedly on $\overline{v}(M,M^1)$. It appears to be possible to fit the observational data by choosing appropriate $\overline{v}(M,M^1)$, but it is not shown that the choice will be realistic and that the theory can therefore be regarded as relevant in practice.

A more detailed study of the effects of cloud coalescence has been made by Taff and Savedoff (1973). They consider cases with and without equipartition of the kinetic energy of translatory motions of clouds, and examine total coalescence and partial coalescence with some of the mass of the colliding clouds going free.

Like Nakano they find that an equilibrium situation is reached fairly quickly and is independent of the initial mass spectrum.

The results of numerical calculations are presented in the form of values of Y in the formula

$$N(m) \propto m^{-Y} \qquad (11.22)$$

where N(m) is the number of objects in unit range of mass at m.

With total coalescence, $1.3 < Y < 1.8$ depending upon the distribution of velocities and cross sections and whether or not there is equipartition of kinetic energy (with equipartition,

Y is about 0.2 less than without it). If there is only partial coalescence the result depends critically on whether the mass not coalescing is in one piece or two. In the first case $Y \approx 1.6$; in the second $Y \simeq \infty$, that is to say all the mass goes into small fragments. Clearly, therefore, one can obtain any value of Y between 1.3 and infinity by choosing particular relative probabilities of the different types of coalescence. In such circumstances it is difficult to test the theory by comparison with observation.

11.04 Fragmentation Induced by Molecule Formation

The data concerning the locations of star formation and the spatial relationship between young stars and interstellar matter, chapter 2, suggest that stars form in the most quiescent cold, dark clouds containing abundant molecules. Star formation appears to be a cause rather than a consequence of turbulence, but the evidence suggests that it could in some way be a consequence of molecule formation.

The possibility that the formation of H_2 molecules may give rise to pressure instabilities leading to cloud fragmentation and star formation was suggested by Schatzman (1958). He discussed the effects on the pressure of the reduction in number density by a factor two and the enhanced rate of cooling which would follow molecule formation. A larger reduction in number density due to the condensation of substantial fractions of the interstellar hydrogen on to interstellar grains has also been suggested (Reddish, 1968a) but it is not certain that the grains can become cold enough for this to occur.

Although compression by turbulence may cause initial density increases in interstellar clouds, it is here suggested that it is the formation of molecules which causes fragmentation of clouds and the formation of stars. In what follows it will be assumed that the cloud density has been increased through compression by turbulence and density waves to a value at which molecule formation begins.

It is generally accepted that much if not all interstellar molecule formation takes place on the surfaces of the interstellar grains (de Jong, 1972) although alternative situations with molecules forming by charge exchange processes (Yoneyama, 1972) or on planetisimals (Horedt, 1972) have been examined. The rate of formation dN/dt is proportional to the residence time of the appropriate atoms on the grains which depends sensitively on grain temperature, so that at grain temperature T

$$dN/dt \propto e^{E/kT} \tag{11.23}$$

where E is the physical adsorption energy of the atoms on the grains. For example, in the case of H_2 molecules forming on grains of ice and such like (Lee, 1972), E/k is 860K and the critical grain temperature is 20K; consequently a change in grain temperature by 1% changes the rate of formation of H_2 molecules by 53%. More complex interstellar molecules such as OH, H_2O etc., have higher adsorption energies and form at rather high temperatures at rates similarly temperature sensitive.

A coating of molecules greatly enhances the far infrared emissive power of grains, causing them to cool rapidly. This in turn causes an increased rate of molecule formation on grain surfaces. In a cloud of uniform density and grain temperature this would result in nearly all of the hydrogen combining to molecules, but in practice a cloud will not be uniform and an interesting situation develops as follows.

Variations in density will occur within a cloud superimposed on any general radial density gradient and these, together with irregularities in shape, will cause the optical depth due to grains to fluctuate irregularly through the cloud, as well as increasing towards the cloud centre. Since the grains heat and cool by radiation this will cause corresponding fluctuations in grain temperature and in the local rates of formation of molecules.

Density variations will generally occur over all scales of

distance within the cloud. Over distances corresponding to optical depth $\tau << 1$ they will not significantly affect the optical depth from the surface at any point. Over distances corresponding to $\tau >> 1$ they will cause systematic variations such as from the edge of the cloud to the centre; but over distances corresponding to $\tau \sim 1$ they will be the cause of significant local differences in optical depth at a given distance from the surface.

A grain will "see" other grains up to a distance corresponding to about unit optical depth. All the grains within this distance are subject to essentially the same radiation field and take up a fairly common temperature. Over distances greater than unit optical depth the grains cannot communicate directly with each other, they are subject to different radiation fields and may take up different temperatures. Unit optical depth is the scale length of distances over which grain temperatures vary irregularly.

It follows that when molecule formation takes place in an interstellar cloud the cloud will take on a cellular structure. In regions where grains are slightly colder than average molecule formation proceeds faster, and the grains cool faster due to enhanced radiation by molecule-coated surfaces. The radii of these cells of molecules will correspond to unit optical depth.

This conclusion is not affected by the likelihood that the temperature may be lower and the density higher in the centre of the cloud, so that molecule formation begins there. As it spreads outwards, it first penetrates the colder cells in a region and the cellular pattern is inevitably impressed upon it.

Within the cells, the combination of hydrogen to molecular form reduces the number density by a factor two, and the enhanced emissive power of both gas and grains due to molecule formation causes the temperatures of gas and grains to fall markedly. Measurements of excitation temperatures in dark clouds support theoretical expectations that the final temperature is probably 5K or less (Schatzman, 1958; de Jong, 1972). The result is a drop in the internal pressure in the cells by an order of magnitude or more. If this takes place in a time less than the gravitational contraction time of the cloud the cells will be compressed by the intercellular gas which, becoming less dense and consequently hotter through diminished radiative power, will be even less able to combine to form molecules. Thus dense globules of molecules condense in an ambient gas of atomic hydrogen.

It does not matter how small the initial density and temperature variations are; the situation is unstable due to the positive feedback inherent in the process "grains cool → molecules form → grain radiation enhanced." The development of a cellular pressure structure and subsequent fragmentation is inevitable.

For example, consider a typical fairly dense interstellar cloud of 10^3 solar masses and density $n_H = 4000\ cm^{-3}$. The cloud will have a diameter of 3 pc and, with grains in normal abundance $n_g/n_H = 10^{-12}$ having optical cross sections $10^{-9}\ cm^2$, the optical depth to the centre will be $\tau = 20$. In these conditions H_2 molecules will form on grain surfaces (de Jong, 1972; Lee, 1972).

Density variations by 5% will cause the optical depth at $\tau = 5$ to vary by $\pm$ 0.1 over transverse distances of $\tau = 1$. Now in such a cloud the grain temperature varies with optical depth (Werner and Salpeter, 1970) as

$$dT/(grain)/d\tau \approx -1 \qquad (11.24)$$

and consequently the grain temperature will vary by $\pm$ 0.1K also over distances corresponding to unit optical depth. Such variations are superimposed on any systematic temperature gradient through the cloud. They will cause corresponding differences in the rate of molecule formation which, at the grain temperature (10K) at this depth will be by a factor 2.3. Changes in the rate of molecule formation due directly to the density variations will be much less - only about 10%.

11.041 The Contraction of Cloud Fragments

The radius r of a typical cell of the gas cloud, at the time fragmentation begins and prior to the subsequent contractions of individual fragments, corresponds to unit optical depth. Thus if a is the radius of interstellar grains having extinction coefficients Q and number density n_g, we have

$$\pi a^2 Q n_g r = \tau \approx 1 \tag{11.25}$$

and hence

$$r = (\pi a^2 Q n_g)^{-1}. \tag{11.26}$$

The mass of hydrogen in the cell is then

$$M = \frac{4}{3} \pi m_H (\pi a^2 Q)^{-3} \left(\frac{n_g}{n_H}\right)^{-3} n_H^{-2} \tag{11.27}$$

or

$$M = 3.5 \times 10^{-57} (\pi a^2 Q)^{-3} \left(\frac{n_g}{n_H}\right)^{-3} n_H^{-2} M_\odot \tag{11.28}$$

where n_H is the number density of hydrogen atoms immediately prior to combination to H_2 molecules. For simplicity the additional mass of helium will be neglected and equation (11.28) will be taken to represent the total mass of the cell.

If this cell is to become a fragment of the cloud, the time scale for molecules to form in it must be less than the time for gravitational contraction of the cloud as a whole; the fall in pressure in the cell will then cause it to be compressed more quickly than the average separation of cells decreases due to contraction of the whole cloud.

The time scale for gravitational contraction of a cloud of density ρ is

$$t(G) \approx (4\pi G\rho)^{-\frac{1}{2}} = 8.4 \times 10^{14}\ n_H^{-\frac{1}{2}}\ s \qquad (11.29)$$

Recent laboratory experiments on the formation of H_2 molecules on cold surfaces in ultra high vacua have given much improved basic data for the process and have considerably changed our earlier understanding of it. Reliable measurements of the physical adsorption energy of H_2 on ice have led to the result (Lee, 1972) that interstellar H_2 molecules will form on grains without enhanced binding sites, at grain temperatures of 20K, a substantially higher temperature than previously deduced from erroneously low physical adsorption energies. Furthermore, measurements of the rate of transfer of recombination energy to the cold surface (Marenco et al, 1972) have shown that it increases with the coverage of molecules on the surface. It follows that either the rate of formation of molecules is enhanced by greater coverage, or the transfer of energy to the surface is made easier. However, the latter in itself implies more efficient conditions for surface recombination, and consequently it is concluded that the rate of formation of molecules increases with surface coverage of molecules. The experiments indicate a direct proportionality over a considerable density range. Thus the equation governing the rate of formation of interstellar molecules differs from earlier formulations by the addition of a term in n_{H_2}. Thus in the usual equation

$$\frac{dn_{H_2}}{dt} = \gamma\pi a^2 v_H n_g n_H$$

where γ is the rate coefficient, the experiments indicate $\gamma \propto n_{H_2}$ over a wide range so that the equation becomes

$$\frac{dn_{H_2}}{dt} = \gamma_o \pi a^2\ V_H n_g n_H n_{H_2} \qquad (11.30)$$

The maximum value of $\gamma = \gamma_o n_{H_2}$ is unity; the maximum possible value of equation (11.30) for any given n_g and n_H would be if $\gamma_o = 1$ when $n_{H_2} = 1$. In that case, with $v_H = 1.2 \times 10^5$ cm s^{-1} corresponding to T = 60K as indicated by HI 21 cm line radiation measurements in average dark clouds, the time scale $t(H_2)$ for molecule formation, given by $dn_{H_2} \sim n_{H_2}$, would be

$$t(H_2) = 8.2 \times 10^{-6} (\pi a^2)^{-1} \left(\frac{n_g}{n_H}\right)^{-1} n_H^{-2} \text{ s.} \tag{11.31}$$

For fragmentation, $t(H_2)$ must be less than $t(G)$; that is, from equation (11.29) and equation (11.31),

$$n_H > 4.5 \times 10^{-14} Q^{2/3} (\pi a^2 Q)^{-2/3} \left(\frac{n_g}{n_H}\right)^{-2/3} \text{cm}^{-3} \tag{11.32}$$

The right hand side of this inequality will be greater if γ_o has not reached unity by the time that $n_{H_2} = 1$ but the observational data (chapters 2 and 7) does indicate that H_2 molecules are abundant when the average gas density is not much higher than the average value, which is of order unity.

Condition (11.32) applied to equation (11.28) gives an upper limit to the mass of a cloud fragment;

$$M < 1.7 \times 10^{-30} Q^{-4/3} (\pi a^2 Q)^{-5/3} \left(\frac{n_g}{n_H}\right)^{-5/3} M_\odot \tag{11.33}$$

There is no general agreement about the composition of interstellar grains, and about their sizes, extinction coefficients and number density. However, the term

$$(\pi a^2 Q) \, n_g/n_H \tag{11.34}$$

occurring in formula (11.33) is the ratio of obscuration by grains to the number density of hydrogen atoms, and can be written:

$$\pi a^2 Q n_g / n_H = A m_H / \sigma \qquad (11.35)$$

where σ is the column density of hydrogen. This has been well determined on the local region of the Galaxy at the present time to be (Lambrecht and Schmidt, 1957)

$$A_V / \sigma_H = 295 \text{ magnitudes } g^{-1} \text{ cm}^{-2} \qquad (11.36)$$

and hence

$$\pi a^2 Q n_g / n_H = 5.0 \times 10^{-22} \text{ cm}^2 \qquad (11.37)$$

Strictly, $\pi a^2 Q$ should be that for the wavelength region in which the radiation falling on a grain most sensitively affects its temperature. This varies with depth in a cloud, from the near ultraviolet and visible to the near infrared, but since the grain temperature varies as

$$T_g \sim (F(\lambda))^{0.2} \qquad (11.38)$$

it is insensitive to moderate changes in the spectral distribution, and the obscuration in the visible region of the spectrum is consequently a reasonable average to adopt (Werner and Salpeter, 1970).

The grain abundance n_g/n_H may change with time and it may differ from place to place in a galaxy or from one galaxy to another. Consequently it will be useful to separate n_g/n_H from the optical cross section ($\pi a^2 Q$). Since the terms generally occur together in the equations in the form of equation (11.37) no particular significance attaches to the grain model chosen to achieve the separation and the results will not be significantly dependent on the choice. Therefore a model intermediate between graphite, silicates and ice (Wickramasinghe, 1973a, 1973b) has been chosen in which

$$\left.\begin{aligned} \pi a^2 &= 2.5 \times 10^{-10} \text{ cm}^2 \\ Q &= 2 \\ \text{and } \left(\frac{n_g}{n_H}\right) &= 10^{-12} \end{aligned}\right\} \qquad (11.39)$$

at the present time, to accord with equation (11.37).

Using these values for $\pi a^2 Q$ the minimum density a cloud must have for fragmentation to take place, inequality (11.32), may be re-written in terms of grain abundance alone,

$$n_H > 10^{-7} \left(\frac{n_g}{n_H}\right)^{-2/3} \text{cm}^{-3} \qquad (11.40)$$

This is shown as line (a) on Fig. 11.03; fragmentation can take place only in clouds with densities above it. At the present time in the local region of our Galaxy, where $(n_g/n_H) = 10^{-12}$, this implies a density $n_H > 11 \text{ cm}^{-3}$ for fragmentation. Line (a) and this density will be higher if the time scale for molecule formation is greater than the minimum time given by equation (11.31); observational data summarised in chapter 7 indicates that the hydrogen has mostly combined when the density $n_H > 100$ and so it is unlikely that the right hand sides of formulae (11.31) and (11.40) are too low by more than one order of magnitude. This will not significantly affect the results and arguments which follow.

The corresponding maximum mass of a fragment is obtained from (11.33) which using the values (11.39), becomes

$$M < 2.2 \times 10^{-15} \left(\frac{n_g}{n_H}\right)^{-5/3} M_\odot \qquad (11.41)$$

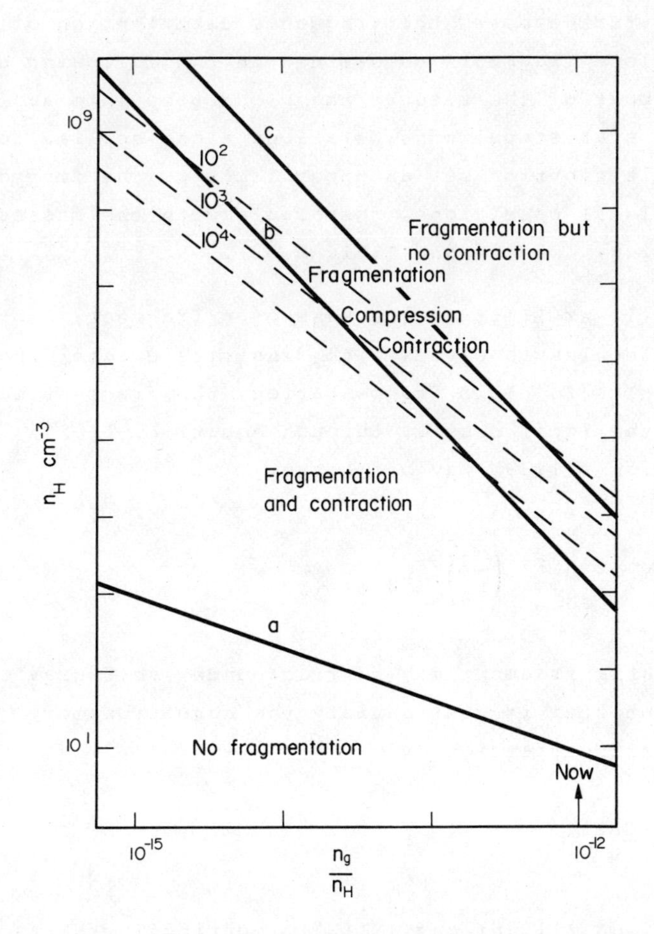

Fig. 11.03
The fragmentation of an interstellar cloud and the contraction of the fragments, in the gas density/grain abundance plane, when fragmentation is induced by the formation of molecular hydrogen on grain surfaces.

Again with $(n_g/n_H) = 10^{-12}$, this gives $M < 2 \times 10^5$ M⊙ as the maximum fragment mass. This high upper limit will not generally be approached because Galactic interstellar clouds do not now have masses large enough for such massive fragments to form from them; furthermore, the frequency distribution of the masses of cloud fragments determined in the following section shows that most of the mass of the cloud goes into small fragments; and statistical considerations alone applied to the frequency distribution set an upper limit to the fragment mass, in present local conditions, comparable with the masses of OB stars (Reddish, 1967b; and Table 3.02).

There is no lower limit to the mass of a fragment; formation of hydrogen molecules in clouds with densities greater than (11.40) results in cloud fragmentation, the fragment mass being related to the local density through equation (11.27) using equation (11.39) may be written

$$M = 2.8 \times 10^{-29} \left(\frac{n_g}{n_H}\right)^{-3} n_H^{-2} \text{ M}_\odot \qquad (11.42)$$

In order that a fragment may contract under self gravitation to become a protostar it must satisfy the condition for gravitational contraction, viz:

$$\frac{3}{5}\frac{GM}{r} > \frac{3RT}{\mu} \qquad (11.43)$$

Using equations (11.26) and (11.27), putting T = 5K as indicated by cold clouds rich in molecules, and with $\mu = 2.3$, this gives

$$n_H < 2.1 \times 10^{-21} \left(\frac{n_g}{n_H}\right)^{-2} \text{ cm}^{-3} \qquad (11.44)$$

as the upper density for contraction. The corresponding fragment mass is

$$M > 6.3 \times 10^{12} \left(\frac{n_g}{n_H}\right) \quad M_\odot \tag{11.45}$$

For example, with $(n_g/n_H) = 10^{-12}$, these give $n_H < 2.1 \times 10^3$ cm^{-3} and $M > 6.3$ M$_\odot$.

At higher densities the cloud breaks into fragments with masses too low to contract under the self-attraction of their own gravitation. It does not follow, however, that these are the smallest fragment masses which can become stars, because a new situation develops when the protostar fragments with masses > 6.3 M$_\odot$ have contracted sufficiently to begin emitting significant amounts of thermal energy. The radiation from them will dissociate molecular hydrogen in their vicinity, within ~ ½ pc (about unit optical depth at the densities involved) and as a consequence the gas pressure there will rise by a factor 2. The time scale for dissociation, < 10^{10}s, is much less than the time for a sound wave to travel that distance and rapid expansion of the dissociated regions will take place, fronted by shock waves. Regions between and ahead of these shock fronts will be compressed.

Now in a typical interstellar cloud in which the density decreases outwards, the fragment mass will increase outwards, as equation (11.42) shows. Consequently the fragments with insufficient mass to contract gravitationally will be towards the cloud centre, and the more massive fragments M > 6.3 M$_\odot$ which become stars will be towards the outside. The fragments inside of these will be compressed by the inwardly advancing pressure wave and then, having smaller radii at a given mass, will satisfy condition (11.43) and contract to form stars. In due course the most massive protostars will become OB stars and will ionise the remaining gas but they will do so unevenly; the more dense and opaque cloud fragments composed of dust and H_2 molecules will not be affected so quickly as the less dense and more transparent HI regions between them: these will expand

and further compress the fragments. The cellular pattern of fragmentation, and the mass spectrum of the fragments will have been imposed on the cloud at the time the H_2 molecules formed.

An increase in n by a factor two due to dissociation and in T from 5K to 10^4 K due to ionisation of HI to HII, will increase the ambient pressure nT by 4 x 10^3 and compress the fragments to radii smaller by a factor 16; by condition (11.43), masses smaller by the same factor will be able to contract to become stars. Consequently, the lower limit to the masses of proto-stars will be reduced from 6.3 M☉ to 0.4 M☉ so that equation (11.45) becomes

$$M \geqslant 4 \times 10^{11} \ (n_g/n_H) \ M_\odot \qquad (11.46)$$

Such a process of compression of cloud fragments, globules or protostars by ionisation of the cloud in which they are embedded, has been fairly widely studied (Dyson, 1973) and there are many photographs of examples; all that has been done here is to determine the process by which the fragments may have come into being and to derive the masses of them.

The situation described appears to be in satisfactory agreement with observational data. The formation of a group of massive stars, M > 6 M☉, followed by that of a dense concentration of dwarf stars 6 M☉ > M ⩾ 0.4 M☉ in its centre, is in accordance with observations showing associations of OB stars typically with dense T-associations embedded in them (Kholopov, 1958; 1960a; 1960b; see chapter 2).

The density given by condition (11.44) is shown as line (b) on Fig. 11.03. Fragmentation takes place at all densities above line (a) when H_2 molecules form; between (a) and (b) the fragments have sufficient gravitational self-attraction to contract to become stars. Above (b) additional compression is required through dissociation and ionisation of the ambient gas by the stars formed between (a) and (b) before contraction to stars can take place; the density limit to which such compression compression is effective is shown as the line (c).

A further limitation must be taken into account, as follows.

Once a cloud has contracted to a density above line (a) it will fragment when the temperatures of the grains in it fall to ~ 20 K so that H_2 molecules form; the rate at which grains cool through this critical temperature is controlled on a cosmic scale by the negative feedback loop (12.02). Calculation and observation both indicate that H_2 molecules form abundantly in any cloud in our Galaxy when it has contracted to a density such that the optical depth through the cloud is ~ 20 (see chapter 7). It seems unlikely that clouds will contract to much higher densities without forming molecules and hence fragmenting. There is of course another negative feedback in this situation, envisaged by Fowler and Hoyle and referred to in chapter 12. If the rate of star formation increases, more turbulent energy is fed into the gas clouds which in consequence are expanded to lower density (and lower opacity) through having a higher internal turbulent pressure.

Using equations (11.25) and (11.42), lines of constant mass for clouds of uniform density having radii corresponding to optical depths of 20 are drawn on Fig. 11.02.

Coalescence of clouds, increased by the 'sweeping up' effect of spiral arms, and the two density equilibrium state of the interstellar gas described in chapter 8, result in low mass clouds generally being too hot to contract to high opacities before they are swept up to become parts of larger aggregates. It appears unlikely that the physical and dynamical conditions in the interstellar gas will permit clouds with masses below ~ 10^3 M⊙ to contract directly to high opacities. It follows (Fig. 11.03) that the line marked 10^3 M⊙ forms an effective upper limit to the density at which fragmentation takes place, giving a lower limit to the stellar masses of 0.12 M⊙. Since this discussion has been in terms of average values, this should perhaps be regarded as a lower limit to the average stellar mass, so that M_1 would become 0.025 M⊙ until meeting line (c) at $(n_g/n_H) = 10^{-13}$. Thus,

$$M_1 \nless 0.025\ M_\odot \tag{11.47}$$

becomes an additional condition to (11.46). It is not necessarily a rigorous limitation; small clouds may sometimes contract to very high densities and fragment into stars of lower masses; and it may differ somewhat from galaxy to galaxy, within a galaxy, or with time; but it is apparent that there will be an effective limit at about this level on most of the mass condensing into stars.

The above discussion has related primarily to fragmentation and the formation of stars in gas with a grain abundance $(n_g/n_H) = 10^{-12}$ appropriate to present local conditions in the Galaxy. Fig. 11.02 shows that if the grain abundance is lower, as may have been the case in earlier epochs, fragmentation and star formation would occur over a wider range of cloud densities and a wider range of protostar masses would result (equation (11.42)). Some interpretations of the consequences of such differences, which are considered in chapter 13, are noted on Fig. 13.01.

Since it follows from the frequency distribution of protostar masses derived in the next section that the average stellar mass is about four times the minimum, it follows further that the average protostellar mass is related to grain abundance by

$$\bar{M}(\text{protostar}) \sim 1.6 \times 10^{12} \left(\frac{n_g}{n_H}\right)\ M_\odot \tag{11.48}$$

11.042 The Initial Mass Function of the Protostars

The frequency distribution of the masses of stars at the time of formation, the initial mass function, will be denoted here by

F(M), defined as the fraction of the number of stars formed with masses in unit mass range at M, so that

$$F(M) = \frac{N(M)}{\int_{M_1}^{M_2} N(M)\,dM} \quad ; \quad \int_{M_1}^{M_2} F(M)\,dM = 1, \tag{11.49}$$

where M_1 and M_2 are lower and upper mass limits, as may be given by (11.46) and (11.41).

Consider an interstellar cloud at the moment at which fragmentation begins as described in section 11.08. The cloud will have a uniform grain abundance, that is a given value of n_g/n_H, but the cloud density will increase towards the centre. By equation (11.28) the masses of the fragments will decrease towards the centre. Given the dependence of density on radial distance within the cloud the fragment masses can be obtained from equation (11.28) and the numbers of them calculated from the known amounts of mass at each density, enabling the initial mass function to be determined. For simplicity only the mass of hydrogen will be considered. Then

$$\rho(r) = m_H \; n_H(r) \tag{11.50}$$

where m_H is the mass of the hydrogen atom, and from equation (11.28) the mass of a fragment formed in gas of density ρ at distance r from the cloud centre is

$$M = \text{constant.} \left(\frac{n_g}{n_H}\right)^{-3} \rho^{-2} \quad M_\odot \tag{11.51}$$

The number of such fragments in the mass range δM at M is

$$N(M)\;\delta M = 4\pi r^2 \rho \;\delta r/M \tag{11.52}$$

where from equation (11.51)

$$\delta M = -\text{ constant}\left(\frac{n_g}{n_H}\right)^{-3} \rho^{-3}\frac{d\rho(r)}{dr}\,\delta r \tag{11.53}$$

Equations (11.51), (11.52) and (11.53) give

$$N(M) = -\text{ constant}\left(\frac{n_g}{n_H}\right)^{-3} M^{-3}/\frac{1}{r^2}\frac{d\rho}{dr} \tag{11.54}$$

or, for any given (n_g/n_H),

$$\log N(M) = \text{constant } -3\log M - \log\left(\frac{1}{r^2}\frac{d\rho}{dr}\right) \tag{11.55}$$

In order to examine the N(M) predicted by equation (11.55) for interstellar clouds, it is convenient to represent the density distributions in the clouds by polytropes. For a polytrope of index m,

$$\rho \propto u^m \tag{11.56}$$

where u is a dimensionless variable representing temperature or gravitational potential energy. Normalised values of $\rho(r)$ have been tabulated for various m (British Association for the Advancement of Science, Mathematical Tables, Vol. II, Emden Functions, London, 1932). They enable $\frac{1}{r^2}\frac{d\rho}{dr}$ to be calculated as functions of ρ and hence, by using equation (11.51) as functions of M, for various m. In this way N(M) has been determined for various polytropes. The results are shown in Fig. 11.01 where they are compared to the observational data (Salpeter, 1955) with which there is good agreement. The various polytropes represent varying degrees of central concentration of mass in the clouds; curves are shown for indices $m = \infty$ (isothermal), 3, 2 and 1.5 (convective) for which the ratios of central to mean density are $\rho_c/\rho = \infty$, 54, 11 and 6 respectively. It is notable that the calculated initial mass functions for this wide range of central concentrations do not differ greatly and neatly straggle the observational data.

11.05 Other Theories

A number of other theories have been advanced to account for the phenomenon of fragmentation of interstellar clouds, without predicting a frequency distribution of fragment masses to enable the crucial comparison with the observations to be made.

Shock wave interactions are examined by Grzedzielski (1966), who finds that the successive compressions produced by the passage of three or more shock waves through a cloud of pregalactic mass and density cause local density increases by factors as large as several hundred and lead to large scale fragmentation; it is possible that a similar process may operate on the scale of interstellar clouds leading to the formation of protostars.

A cloud of neutral gas embedded in a less dense ionised gas will be compressed by the propagation into the cloud of the shock wave resulting from the pressure difference across the ionisation front. If the shock wave is strongly cooled, the degree of compression will be considerable and may cause the compressed region immediately behind the shock to become unstable to gravitational contraction for masses of stellar size. The conditions necessary for instability have been determined by Kossacki (1968) who found that they are an initial density of 10^3 atoms cm^{-3} and a temperature behind the shock of 40 K.

The presence of a magnetic field of sufficient strength would, if it were linked to the gas by a sufficient degree of ionisation, inhibit contraction of a cloud across the lines of force; if the cloud is rotating the axis of rotation would be induced to lie along the lines of force, and the cloud would then contract to a disc. The fragmentation of such a disc has been considered by several authors, recently including Disney et al, (1969), Larson (1972a) and Nakano (1973). That a disc would fragment appears to be well established, but it is not established that fragmentation would not occur before contraction to a disc, in part because the initial conditions of rotation and magnetic field strength are not known.

11.06 Summary

Of the various types of theories which have sought to account for the formation of stars, only two appear to remain as serious contenders. In both of them it is supposed that stars form as a result of the fragmentation of interstellar clouds and the contraction of the fragments to stellar densities, the fragmentation and contraction being caused by pressure and density variations in the cloud. In the first type, the variations result directly from turbulence; in the second they result from the combination of atoms to form molecules or solids.

Those theories which depend on turbulence suffer, however, from three substantial weaknesses that have prevented them from forming a convincing explanation. The first is that turbulence appears to be unable to produce density variations large enough to cause the instabilities required for fragmentation and self-contraction of fragments of stellar mass (Grzedzielski, 1966; Roberts, 1969; Kalnajs, 1970). Secondly, there is no observational data which shows that stars form in turbulent clouds - on the contrary newly formed stars appear to occur in regions associated with the most quiescent, cold, dark clouds containing abundant molecules (Mezger et al, 1967, 1968; chapter 7); star formation appears to be a cause rather than a consequence of turbulence in clouds. Thirdly, the theories have not succeeded in giving quantitative results to compare with the major features of the relevant observational data, such as the frequency distribution of stellar masses and the rate of star formation, without the introduction of arbitrarily chosen parameters.

On the other hand, the formation of molecular hydrogen in clouds is shown to induce pressure instabilities leading to cloud fragmentation and the formation of protostars with a mass function in good accord with observational data - better in fact than for any of the alternative theories - and unlike other theories having no arbitrary parameters which can be adjusted to produce agreement between theory and observations. It is perhaps worth emphasising that the corresponding mass function calculated in section 11.07 and shown on Fig. 11.01 is the inevitable consequence of the laws of physics applied to a contracting cloud of interstellar hydrogen containing some solid grains.

The need to account for the observed initial mass function is perhaps the most crucial test of any theory of star formation. On the basis of the comparisons between theory and observation detailed in the previous sections of this chapter and shown on Fig. 11.01 it appears that star formation results from the formation of H_2 molecules or solids in clouds.

Chapter 12.

THE RATE OF FORMATION OF STARS.

Few of the theories of star formation discussed in earlier chapters consider the rate at which stars form. There appears to be a widespread view, usually implicit rather than explicitly expressed, that it is a statistical rather than a physical problem; that the actual rate is the rather arbitrary result of a chance combination of the forces acting to assist and to resist star formation. The writer doubts that so fundamental a property of the evolution of the Cosmos would have been left to chance.

Physical theories involve the operation of negative feedback loops to control the rate, and two such theories have been put forward.

In that by Fowler and Hoyle (1963), control operates through the loop

$$\text{clouds contract} \rightarrow \text{stars form} \rightarrow \text{radiate energy} \rightarrow \text{increase turbulence} \rightarrow \text{clouds expand} \rightarrow \text{turbulence decays} \rightarrow \tag{12.01}$$

The rate of star formation is determined by the rate at which the turbulent energy is dissipated, and the control is essentially of a local character. In the theories by Reddish and Wickramasinghe (1969) and by Reddish (1975) the control is by the loop

$$\text{grains cool} \rightarrow \text{molecules combine} \rightarrow \text{clouds fragment} \rightarrow \text{stars form} \rightarrow \text{heat grains} \rightarrow \text{cosmic expansion} \rightarrow \tag{12.02}$$

and the control is largely cosmic since the radiation field is predominantly universal. Both of these theories were described in chapter 11 in connection with the initial mass function, and will now be considered in relation to the rate of star formation.

12.01 Gravitational Turbulence

Fowler and Hoyle base their theory on what they call 'gravitational turbulence'. In this, the physical conditions are such that an increase in density increases the self gravitation and the ability of a cloud to contract, and they contrast this with pressure turbulence in which compression is accompanied by a rise in internal pressure tending towards re-expansion. The former assists star formation, the latter opposes it. The energy radiated by the stars creates the pressure turbulence which increases the internal energy of clouds, opposing their contraction to create more stars. Thus the negative feedback loop is formed and the rate of star formation is controlled by the rate at which the pressure turbulence dissipates. This is estimated to be 10^{-27} erg cm^{-3} s^{-1}, and for equilibrium about 1% of the energy radiated by recently formed stars, ~ 10^{-25} erg cm^{-3} s^{-1}, would have to be converted into turbulent energy in the interstellar gas.

The manner in which turbulent energy is fed into interstellar clouds by the thermal radiation from stars has been examined in some detail by Field and Saslaw (1965), following earlier suggestions by Oort (see chapter 8). The formation of stars, including massive O-stars, in a large cloud ionises the gas in their vicinity and the expanding HII region thus formed compresses and disrupts the surrounding gas, thereby ejecting small dense clouds into the interstellar medium. The individual clouds in the general distribution formed by such events and distributed throughout the galaxy collide and coalesce, building up larger aggregates which in time become gravitationally unstable and contract to form stars, thus continuing the cycle:

$$\text{large clouds contract} \rightarrow \text{stars form} \rightarrow \text{large cloud disrupted} \rightarrow \text{clouds collide and coalesce.} \rightarrow \qquad (12.03)$$

The rate at which matter passes through this cycle, and the fraction of it which condenses to form stars, then determines the rate of star formation. Field and Saslaw make the following estimate. The masses of unstable clouds in which star formation is occurring are thought to be about 4×10^4 M⊙. The ejected fragments may have masses of about 30 M⊙ each. In order to ionise the remaining cloud, some 3% of the cloud mass must condense into stars. Consequently it is estimated that each aggregate ejects 97% of $4 \times 10^4/30$ or some 1300 small clouds. The model calculations which are made give estimates of the rate at which the small clouds collide and coalesce, and have the time to regenerate the initial situation. The mass formed into stars in a given time is thus determined and found to agree with observations. However, the estimate relies so heavily on empirical values obtained from observation that the agreement is hardly surprising; the most that can be said of the theory is that it is internally consistent. It does not provide, on a physical basis, an independent prediction of the rate of star formation which can be used to test the theory. It is probably capable of doing so although our present understanding of the complexities of the process is inadequate. See also section 8.02.

12.02 Molecule Formation

In the theories by Reddish and Wickramasinghe (1969) and Reddish (1975), the formation of H_2 molecules causes pressure instabilities in clouds, leading to cloud fragmentation and star formation. The rate at which stars form is thus governed by the rate of formation of molecules. This in turn is controlled by the temperatures of the interstellar grains on the surfaces of which the molecules form, and the grain temperatures are themselves determined by the interstellar radiation field, which is largely cosmic. The radiation field is added to by stellar radiation, and leaks away through cosmic expansion. The negative feedback loop (12.02) maintains the equilibrium - stars form in just such numbers that the energy they radiate replaces that lost by cosmic expansion in the same time. Thus the rate of star formation is controlled by the rate of cosmic expansion. If too many stars form, the heat supply exceeds the rate of loss, the grains become hotter, molecule formation is reduced, fewer clouds fragment and fewer stars are formed; and vice-versa. The situation is analogous to a room thermostat controlling a domestic heating system. It was shown in chapter 11 that the loop gain is very high.

An estimate of the average rate of star formation expected on the basis of this theory can be made as follows.

The effective temperature of the cosmic background radiation is 2.7 K and that of the total cosmic radiation field is about 3.0 K. Taking the rate of cosmic expansion to be $H = 2 \times 10^{-18}\ s^{-1}$ the rate of loss of radiant energy is

$$\frac{dE}{dt} = 3aT_e^4 H = 3.7 \times 10^{-30}\ \text{erg cm}^{-3}\ \text{s}^{-1} \qquad (12.04)$$

The mean smoothed-out mass density of stars has been determined by Oort (1958b) to be 3×10^{-31} g cm^{-3}, whereas Schatzmann (1958) obtains 2×10^{-30} g cm^{-3}.

To balance equation (12.04) Oort's value would require an average emission of 12 erg g s^{-1}, a mean mass/luminosity ratio of 0.17 for galaxies. Schatzmann's value would require 1.8 erg g^{-1} s^{-1}, an M/L ratio of 1.1, closer to observation but still probably too low. However, the determination of the mean cosmic mass density of stars could be in error by an order of magnitude.

Another and perhaps more instructive approach is to consider the rate of condensation of mass into stars necessary to maintain the cosmic star density at its present value.

$$\frac{1}{\rho}\frac{d\rho}{dt} = 3H = 6 \times 10^{-18}\ \text{s}^{-1} \qquad (12.05)$$

This would balance only that part of the energy loss in equation (12.04) which corresponds to radiation derived from galaxies of stars. If as appears to be the case, two-thirds of equation (12.04) is due to the cosmic background radiation, equation (12.05) would need to be multiplied by a factor 3 to compensate for that as well. That is, the mass density in stars would need to be increasing to replace, by increased stellar radiation, the diminishing energy density of the cosmic background radiation.

The observed rate of condensation of mass into stars in the local group of galaxies has been found (Reddish, 1962a) to be

$$\frac{1}{M}\frac{dM}{dt} = 1.3 \times 10^{-17}\,\frac{M(\text{gas})}{M}\ \text{s}^{-1} \tag{12.06}$$

where M is the mass in stars. With $M(\text{gas})/M \sim 0.1$, equation (12.06) may be an order of magnitude too small in comparison with equation (12.05). The consequences of using it and other values in calculations of model galaxies is examined in the following chapter.

It is concluded that although the present slow rate of star formation indicates that the average grain temperatures must be close to the critical value for molecule formation, and that consequently the rate of star formation may be controlled by the rate of cosmic expansion operating on the radiation field, existing data are too inaccurate either to confirm or to deny the possibility.

A point of interest is the dependence of the rate of star formation on the effective temperature of the interstellar radiation field, through its effect on grain temperature.

Making the reasonable assumption that the rate of condensation of mass into stars is proportional to the rate of formation of H_2 molecules we have, using equation (11.23),

$$\frac{dM(stars)}{dt} \propto \frac{dN(molecules)}{dt} \propto e^{-860/T_g} \qquad (12.07)$$

where T_g is the grain temperature. The critical grain temperatures for H_2 formation is $T_g \approx 20$ K (Lee, 1972).

The emissive power and absorptivity of grains is not well known, but for a given spectral distribution and for small changes in T_g, we may reasonably expect T_g to vary in direct proportion to the effective temperature of the interstellar radiation field. Proportionality (12.07) then gives the result shown in Fig. 12.01. This illustrates the high gain in the negative feedback loop (12.02) and shows how critically the rate of star formation depends on the ambient radiation field. This result will have a particular significance in the following chapter.

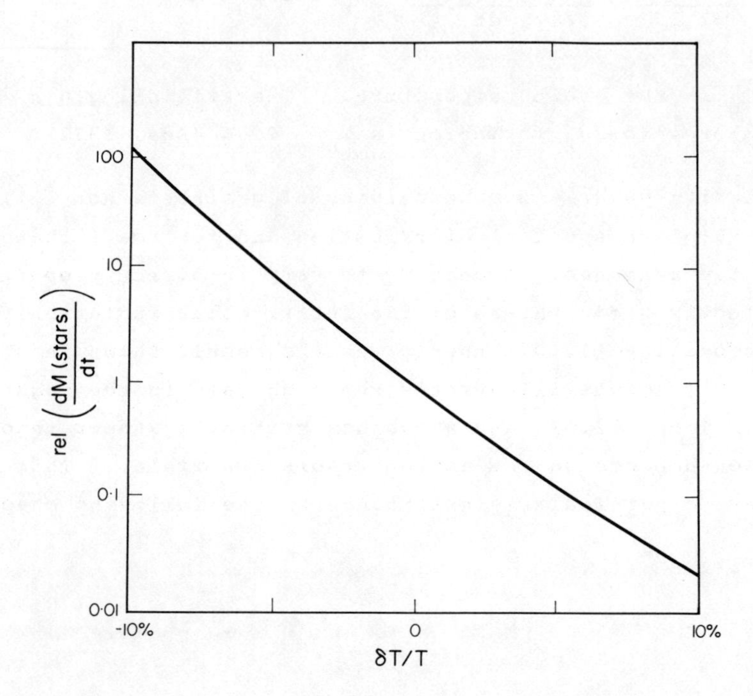

Fig. 12.01
The effect on the rate of star formation of changes in the effective temperature of the interstellar radiation field, when star formation results from cloud fragmentation induced by the formation of interstellar molecules on grain surfaces.

Chapter 13.

THE EVOLUTION OF GALAXIES.

Galaxies are the primary visible components of the Universe at large, and an understanding of their formation and evolution is an essential part of our exploration of the Universe.

Galaxies are assemblies of stars and interstellar matter; the stars are formed from that matter, and eject some of their material back into it. Given a knowledge of the distribution of interstellar matter, of the way in which stars form from it, and of the subsequent evolution of them, it should be possible to describe the appearance of the galaxy in the past, to account for its present state and to predict its evolution in the future.

Much is known about the evolution of stars, the changes in element abundances that take place within them, and the ejection of matter from them, although there are still important deficiencies in our knowledge particularly of the less stable periods of evolution.

The present distribution of interstellar matter is quite well determined from observation, but little is known with any certainty about its change with time because of uncertainties about the forces acting on matter on the large scale.

The extent of our understanding of the processes of star formation may be judged from the earlier chapters of this book. Although tentative, some of it is beginning to be put on a physical basis. So far as it affects calculations of the evolution of galaxies of stars three parts of it are of crucial importance; they are the rate of star formation, the frequency distribution of protostar masses, and the dependence of both of those on the physical and chemical condition of the interstellar matter out of which stars are forming.

Using such knowledge as has been available at the time, many theoretical studies have been made of the evolution of galaxies (Hillel, 1973; Ikeuchi et al, 1972; Larson, 1974; Matsuda, 1970; Reddish, 1960, 1968a, 1969b, 1970, 1975; Reddish and Wickramasinghe, 1969; Searle et al, 1973; Tinsley, 1968, 1972, 1973). Most of these are more or less empirical, only the later ones by Reddish (1969a, 1969b, 1970, 1975) and by Reddish and Wickramasinghe (1969) being based wholly on physical theories of the formation and evolution of stars, theories which are included in part in earlier chapters of this book. Partly on account of their physical basis, but also to maintain a consistent line of argument, they will be used as the foundation of this chapter and the analyses by other authors will be described in relation to them.

13.01 Star Formation in the Proto-galaxy; Controlling Influences of Grains and Molecules

The starting point of our theory is a contracting proto-galactic cloud of atomic hydrogen containing a small abundance of solid grains. As contraction proceeds, a density is reached at which atoms begin to combine to form molecules on the surfaces of the grains, a process which occurs non-uniformly and which results in pressure instabilities, cloud fragmentation and star formation in the manner described in chapter 11. The condition at which this begins is shown as line (a) in Fig. 11.03.

This process of star formation and the subsequent evolution of the galaxy of stars and gas is controlled by a number of feed-back loops, whilst the frequency distribution of the masses of stars formed is determined by physical properties of the grains.

As was shown in chapter 11, cloud fragmentation and _star formation_ is induced by the instability resulting from the positive feedback in the loop.

grains cool → molecules form → gas density and grain radiation increases →

The overall rate of star formation depends on the rate at which H_2 molecules form and is governed by the negative feedback loop (12.02). The determining factors in this loop are the rate of cosmic expansion which gives the rate at which the cosmic heat is diluted, and the frequency distribution of the masses of stars formed, on which depends the heat output of the mass condensed into stars. These control the rate of condensation of mass into stars by determining the constant of proportionality and T_g in (12.07) giving in average conditions the rate equation (12.06). For purposes of studying a single galaxy this is more conveniently given in the form

$$\frac{d(N\bar{M})}{dt} = KM(\text{gas}) \qquad (13.01)$$

where K varies with grain temperature as in Fig. 12.01. In the local group of galaxies, $K = 1.3 \times 10^{-17}\ s^{-1}$, but it will appear from later calculations in this chapter that the average value among galaxies is probably about twice that. The frequency distribution of stellar masses is determined by the opacity of the grains and is given by the equation (11.55). The average of the distribution produced by clouds with a wide range of density concentrations, Fig. 11.01, is well represented by the formula

$$N(M) \propto = CM^{-7/3} \qquad M_1 \leqslant M \leqslant M_2 \qquad (13.02)$$

where C is a constant, and from (11.41)

$$M_2 \leqslant 2.2 \times 10^{-15}\ (n_g/n_H)^{-5/3}\ M_\odot \qquad (13.03)$$

and from (11.46) and (11.47)

$$M_1 = 4 \times 10^{11}\ \left(\frac{n_g}{n_H}\right)\ M_\odot; \quad M_1 \nless 0.025\ M_\odot. \qquad (13.04)$$

The inequality sign occurring in (13.03) becomes relevant when the total mass of gas available is insufficient for such large fragments to form. In practice the upper limit M_2 is given as follows:

When a cloud breaks into fragments, the mass of the cloud is the sum of the masses of the fragments:

$$M(\text{cloud}) = \int_{M_2}^{M_1} N(M)\,M\,dM = C\left[M^{-1/3}\right]_{M_2}^{M_1} . \qquad (13.05)$$

With $M_2 >> M_1$,

$$C = M_1^{\,1/3}\, M(\text{cloud}) \qquad (13.06)$$

The upper limit M_2 is reached when the amount of mass available to form a fragment of mass M_2 is only just sufficient; thus

$$C\left[M^{-1/3}\right]_{\infty}^{M_2} \approx M_2 \qquad (13.07)$$

from which

$$M_2 \approx C^{3/4} \qquad (13.08)$$

and using equation (13.06) this gives

$$M_2 \approx \left(M(\text{cloud})\,M_1^{\,1/3}\right)^{3/4} \qquad (13.09)$$

As noted in chapter 11 perhaps only about half the mass of a cloud goes into fragments, the remaining material remaining as intrafragment gas: so more probably in practice

$$M_2 \approx \left(\frac{1}{2}\, M(\text{cloud})\, M_1^{\,1/3}\right)^{3/4} \qquad (13.10)$$

Using equation (13.04) this becomes

$$\left.\begin{aligned} M_2 &= 470\ (M(\text{cloud}))^{3/4}\left(\frac{n_g}{n_H}\right)^{1/4}\ M_\odot \\ &\ngtr 2.2 \times 10^{-15}\left(\frac{n_g}{n_H}\right)^{-5/3}\ M_\odot \end{aligned}\right\} \qquad (13.11)$$

For example, with $\frac{n_g}{n_H} = 10^{-12}$ as in the local region of our Galaxy at the present time (see chapter 11), M_2 has the values in column 2 of Table 13.01. If at an earlier stage $(n_g/n_H) = 10^{-15}$, say, then M_2 would be as in column 3.

TABLE 13.01

M(cloud)	M_2	
$M_\odot$	$M_\odot$	$M_\odot$
	$(n_g/n_H = 10^{-12})$	$(n_g/n_H = 10^{-15})$
10^2	15	2.7
10^3	83	15
10^4	470	83
10^5	2.6×10^3	470
10^6	1.5×10^4	2.6×10^3
10^8	2.2×10^5	1.5×10^4

TABLE 13.01 Cont.

M(cloud)		M_2
10^{10}	2.2×10^5	4.7×10^5
10^{12}	2.2×10^5	1.5×10^7

The values given by equation (13.11), such as in Table 13.01, only apply when the gas clouds in the proto-galaxy have contracted so that the densities in the centres of them are at least as high as line (c) on Fig. 11.03 at the relevant grain abundance so that the whole permissible range of protostars is being formed. At lower densities, M_1 is given by equation (11.42) instead of by (11.46); thus for example we have the values given in Table 13.02 during the contraction of a proto-galactic cloud through the density range $10^4 \leqslant n_H \leqslant 10^5$ cm^{-3}, when $n_g/n_H = 10^{-15}$.

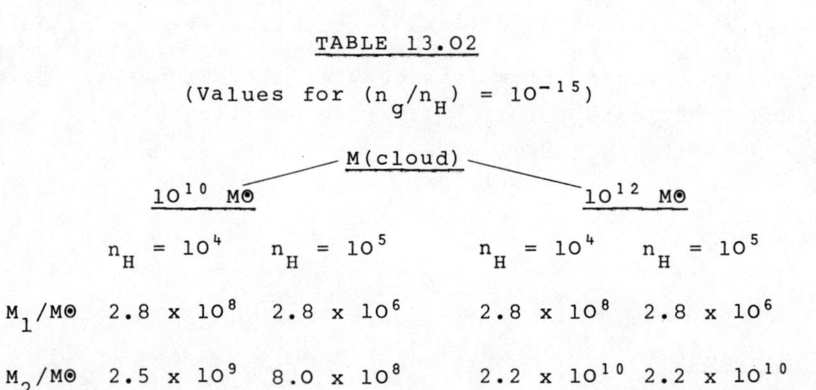

TABLE 13.02

(Values for $(n_g/n_H) = 10^{-15}$)

	M(cloud)			
	10^{10} M⊙		10^{12} M⊙	
	$n_H = 10^4$	$n_H = 10^5$	$n_H = 10^4$	$n_H = 10^5$
M_1/M⊙	2.8×10^8	2.8×10^6	2.8×10^8	2.8×10^6
M_2/M⊙	2.5×10^9	8.0×10^8	2.2×10^{10}	2.2×10^{10}

It follows that supermassive objects form during the contraction phase.

Another negative feedback loop which is of particular significance in this chapter controls M_1 and M_2 through the grain abundance as follows.

The most massive objects first formed in the proto-galaxy are given by relation (13.03) and are proportional to $(n_g/n_H)^{-5/3}$. Typical masses are given in Table 13.02. At a later stage, when the proto-galaxy has contracted so that densities cover the range between conditions (a) and (c) on Fig. 11.03, stars are being formed with masses down to M_1 given by equation (13.04) and the largest fragment masses then forming may be limited by statistical considerations as given by equation (13.11) and in Table 13.01. Now the situation is rather different according to whether or not the mass of the proto-galaxy is greater or less than about 10^{10} M⊙. If it is less, then it can be seen from Table 13.01 that at low grain abundances less massive objects form in it than at higher grain abundances. If the galaxy is greater than 10^{10} M⊙, more massive objects form in it at low than at high grain abundances.

Since the explosion of massive objects (supermassive objects and supernovae) is a source of heavy elements and grains, the rate of production of elements and grains in dwarf and giant galaxies differ somewhat.

a) dwarf galaxies (< 10^{10} M⊙)

low grain abundance
↓
less massive objects (13.12)
↓
lower than average rate of
creation of heavy elements
and grains

and we may expect dwarf galaxies to be relatively deficient in heavy elements (as are globular star clusters).

b) giant galaxies (> 10^{10} M⊙)

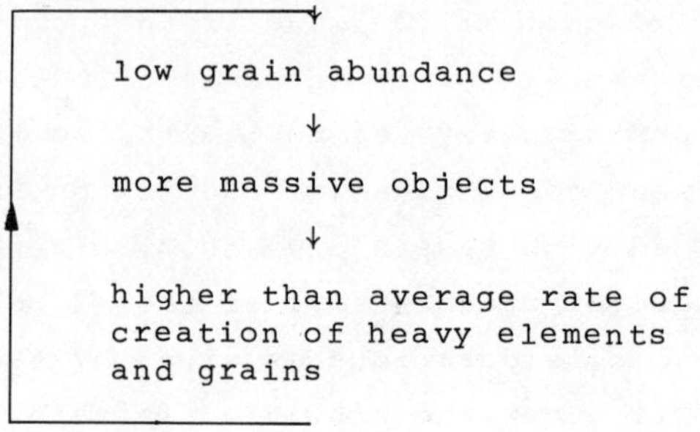

$$\begin{array}{c} \text{low grain abundance} \\ \downarrow \\ \text{more massive objects} \\ \downarrow \\ \text{higher than average rate of creation of heavy elements and grains} \end{array} \tag{13.13}$$

This negative feedback loop will result in all giant galaxies tending in time towards a common abundance of grains and heavy elements rather higher than in dwarf galaxies.

The initial abundance of solid grains is not known, but is presumably considerably less than the present value because grains are produced by stars in the subsequent evolution. The calculations to be described in this chapter were therefore made for various values of $(n_g/n_H)(t = o)$ and it will be shown that the results are insensitive to it. For purposes of illustration results obtained with $(n_g/n_H)(t = o) = 10^{-15}$, which is 10^{-3} of the present value (chapter 11) will be described in detail and the effects of changing the initial value will be shown.

Referring again to Fig. 11.03 the formation of supermassive objects therefore begins when the proto-galaxy has contracted to a density on line (a).

Since the rate of molecule formation at these densities is faster than the rate of free fall contraction (this being the condition for fragmentation represented by line (a)), molecule formation, fragmentation and the condensation of the supermassive objects to a luminous stage will occur before the overall density distribution in the proto-galaxy can change markedly. Thus if the overall density is fairly uniform, supermassive objects will form throughout the galaxy, but if it increases inwards they will form first in the central regions.

In the first case perhaps as much as half of the proto-galactic gas - that where the grains are colder than average - will form molecules and condense into supermassive objects before the first of them become luminous, closing the feedback loop (12.02) locally and inhibiting further molecule formation. Otherwise, the emission of thermal energy by supermassive objects formed first in the central more dense regions will prevent their immediate formation further out by inhibiting molecule formation. In either case further star formation (using "star" in a very general sense) will be slowed down or stopped in the galaxy until the supermassive objects have run through their evolution and ceased to radiate. This will be for a period $\sim 10^7$ yr.

A galaxy may go through a series of such outbursts during its evolution. Averaged over a period of time which is a significant fraction of the Hubble time and over a volume containing many galaxies the rate of condensation of mass into luminous objects must accord with equation (13.01), but the rate for any individual galaxy at any moment may be markedly different. It was pointed out in chapter 12 (Fig. 12.01) that a change by 10% in the effective temperature of the ambient radiation field in which a galaxy finds itself will change the rate of star formation by a factor of a hundred; it will emerge later in this chapter that if such changes occur initially they produce remarkable differences in the overall properties of galaxies at the present time.

It is of interest to consider the size of the proto-galactic cloud when it has contracted so that its density lies on line (a). If n_g/n_H $(t = o) = 10^{-15}$ this gives $n_H = 10^3$ (relation (11.40)) and for a typical galactic mass of $10^{44} - 10^{45}$ g, the diameter of the proto-galaxy at this stage would be 1 to 2 kpc. On the other hand if the initial grain density is as high as $(n_g/n_H)(t = o) = 10^{-13}$ then $n_H = 47$ cm^{-3} and the diameter would be 3 - 6 kpc. In any case the object will be a fairly compact galaxy.

Observations give several reasons for believing that this phase may be correctly identified with the so-called 'compact galaxies'. The luminosity per unit mass is high (O'Connel and Kraft, 1972; Arp et al, 1973), there is a high proportion of very luminous (massive, young) stars (Fairall, 1971), and the interstellar gas content is very high - perhaps most of the total mass (Carozzi et al, 1974; Bottinelli et al, 1973; Lanqué, 1973).

At the end of these outbursts of formation, exploding supermassive objects will have ejected most of their mass into the remaining interstellar gas. The composition of the interstellar mass will now become that of the ejecta of the supermassive objects diluted with the remaining interstellar hydrogen. Consequently the composition of the interstellar gas out of which the next generation of stars will form may have been substantially enriched in helium, heavier elements and grains - by amounts which will be determined below - very early in the evolution of the galaxy. It will be shown later that the extent of this enrichment will have little effect on the physical properties of the galaxy when it has reached the age of 10^{10} yr on account of the operation of the several negative feedback loops, but it would have a marked effect on a galaxy observed early in its evolution.

The emission of energy by supermassive objects will delay contraction of the proto-galaxy to higher densities. The effect of such delay can be examined by comparing the results of evolutionary calculations for a proto-galaxy in free fall with results for a rate of contraction slower by various amounts.

These amounts may be considerable. The free fall contraction time is given by

$$t(G) \sim 1/\sqrt{4\pi G\rho}$$

which, at $n_H = 10^3$ (corresponding to line (a) at $n_g/n_H = 10^{-15}$)

gives $t(G) \sim 10^6$ yr. On the other hand, the time scale for thermal dissipation, whether the energy is in the heat of the gas or in turbulence, is (von Weizsadker, 1951; Schatzmann, 1958)

$$t(T) \sim l/v$$

where l is the dimension of the cloud and v the velocity of sound in the gas.

For example, with

$$l \sim \text{several kiloparsec} \sim 3.10^{21} \text{ cm}$$

and

$$T = 60 \text{ K}; \quad v = 10^5 \text{ cm s}^{-1},$$

we have

$$t(T) \sim 10^9 \text{ yr} \sim 10^3 \times t(G)$$

Since there may be several cycles of heating and cooling; the average rate of contraction of the proto-galactic gas to higher densities may be slowed down by a factor of up to a hundred or more.

At this point it will be useful to summarise before considering the further evolution of the young galaxy.

The proto-galactic cloud, of low density and low grain abundance, is located in the lower left of Fig. 13.01.

It contracts at constant grain abundance, and when the density rises above line (a), molecule formation takes place, resulting in cloud fragmentation and the formation of supermassive objects.

The energy emitted by the supermassive objects heats the remaining gas, slowing down its subsequent contraction, and

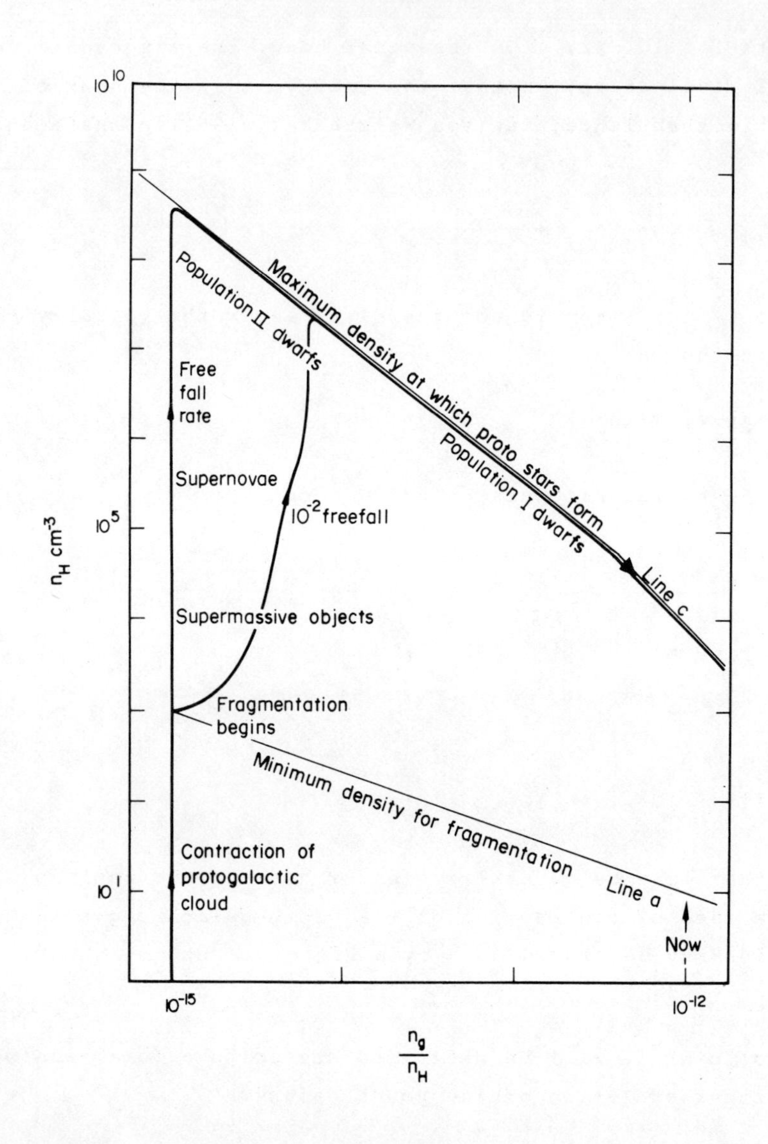

Fig. 13.01

A proto-galactic cloud of low density and initial grain abundance 10^{-15} contracts, and when the density reaches line (a) molecular hydrogen forms on grain surfaces in a shorter time scale than gravitational contraction. Fragmentation results, the masses of the fragments depending on both density and grain abundance. Supermassive objects, supernovae, population II dwarfs and population I dwarfs form in turn, the grain abundance increasing with time due to enrichment of the gas by heavy elements. There is a greater degree of enrichment in population II dwarfs if the rate of contraction of the cloud is slowed much below free fall rate.

their ultimate explosion enriches the gas with heavier elements synthesised in them. Solid grains may also be produced as they are in supernovae explosions, (Seddon, 1969) and the abundances of heavy elements and grains in the gas rises. The upward evolutionary track on Fig. 13.01 begins to curve towards the right.

As some of the gas contracts to higher densities (and hence to higher opacities) less massive fragments form, and the explosion of supermassive objects is followed by the formation of large numbers of massive stars and a consequent outburst of supernovae. These further enrich the galactic gas, and as high densities are reached by the contracting clouds very large numbers of dwarf stars are formed from the enriched material. Thereafter, star formation continues at a declining rate, from progressively more enriched material, over a range of masses narrowing as the boundary limitations (a) and (c) converge.

13.02 Model Calculations

In order to calculate the dependence on time of such quantities as the abundances of heavy elements and grains, the mass of gas uncondensed into stars, the total luminosity, and so forth - all quantities which are required for comparison with observation in order to test the theory - calculations have been made of the properties of model galaxies evolving as described. The parameters of the models are as follows:

Total mass = M_o (13.14)

Abundances by mass fraction: (13.15)

Hydrogen	=	X
Helium	=	Y
Carbon	=	C
All heavy elements	=	Z
Grains	=	n_g/n_H

Initial conditions:

$$M(\text{gas}) = M_o; \quad (n_g/n_H) = 10^{-15}; \quad Z = C = Y = 0; \quad \text{at } t = 0 \tag{13.16}$$

Rate of contraction of gas to higher densities

= R x free fall rate.

Rate of condensation of mass into stars,

K, given by

$$\left.\begin{aligned} \frac{dM(\text{stars})}{dt} &= \frac{dN}{dt}\bar{M} = K\,M(\text{gas}), \\ \text{or} \\ \frac{dN}{dt} &= \frac{K}{M} M_o e^{-Kt}, \end{aligned}\right\} \tag{13.17}$$

where K is controlled by the negative feedback loop (12.02). In the local group of galaxies it is observed to be $\sim 1.3 \times 10^{-17}$ s^{-1} but later calculations indicate that the average value over all galaxies is probably about twice that.

Frequency distribution of stellar masses,

$$N(M) = CM^{-2.33}, \quad M_1 \leqslant M \leqslant M_2 \tag{13.18}$$

where N(M) has been derived in chapter 11 and shown on Fig. 11.01, and M_1 and M_2 are given by equations (13.04) and (13.11) although M_1 is given by equation (11.42) for the densities and grain abundances along the contraction track until the track reaches limit (c).

For material ejected in explosions, the following values were chosen as representing the best estimates based on results of calculations of explosive events and of element synthesis (Hoyle and Fowler, 1960; Colgate, 1968; Wagoner et al, 1967; Hoyle and Wickramasinghe, 1962). SM denotes supermassive objects ($M > 10^3$ M☉), SN denotes supernovae ($M > 10$ M☉) and F the mass of heavy elements (particularly carbon) formed into grains. All quantities refer to mass fractions in the material ejected by the explosions.

$$\left.\begin{array}{l} ZSM = 10^{-5}; \quad CSM = 10^{-4}; \quad YSM = 0.7 \\ ZSN = 0.06; \quad CSN = 0.01; \quad YSN = 0.4 \\ F = 0.04Z \text{ for } M \geqslant 10 \text{ M}_\odot; \\ F = 0 \text{ for } M < 10 \text{ M}_\odot. \end{array}\right\} \quad (13.19)$$

Various properties of this model, as it evolves, can now be calculated.

In the first instance, the calculations are made with a rate of condensation of mass into stars $K = 1.3 \times 10^{-17}$ s^{-1} equal to the observed value in the local group of galaxies, to facilitate comparison with observation, and the consequences of varying the initial grain abundance $(n_g/n_H)(t = o)$, and the rate of contraction of the proto-galactic gas, are examined. Subsequently, the effect of major changes in the initial rate of star formation, as would result from small differences in the initial grain temperatures, are calculated.

13.03 Calculations with $K = 1.3 \times 10^{-17}$ s^{-1}

Since the grains play such a crucial part in the theory of star formation on which the model is based, it is useful to begin with calculations of the growth of grain abundance with time.

13.031 Grain Abundance

From the grain model used in chapter 11, assuming that the grains are composed of material such as graphite or silicates having a density about 2 g cm^{-3}, the mass of a grain is about 1.6×10^{-15} g. The relationship between the mass fraction in grains and the grain abundance is then

$$\frac{\text{mass in grains}}{\text{mass of hydrogen}} = \frac{n_g m_g}{n_H m_H} = 10^9 \frac{n_g}{n_H}$$

Taking the abundance of hydrogen X to be 0.8 on average, we have

$$\frac{n_g}{n_H} = 1.25 \times 10^{-9} \left(\frac{M(\text{grains})}{M(\text{gas})} \right) \tag{13.20}$$

Then

$$\frac{d}{dt}\left(\frac{n_g}{n_H}\right) = 1.25 \times 10^{-9} \frac{d}{dt} \left(\frac{M(\text{grains})}{M(\text{gas})} \right) \tag{13.21}$$

A fairly good approximation is made as follows.

The fraction of mass condensing into stars more massive than M is

$$f(M) = \int_{M}^{M_2} N(M) M dM \Bigg/ \int_{M_1}^{M_2} N(M) M dM \tag{13.22}$$

Putting $M_2 = \infty$, since it is always much greater than M_1, and restricting ourselves to cases where $M_1 < M << M_2$, then using equation (13.18) this becomes

$$f(M) = \left(\frac{M_1}{M} \right)^{1/3}. \tag{13.23}$$

From equation (13.17)

$$\frac{dM(\text{stars})}{dt} = - \frac{dM(\text{gas})}{dt} = 1.3 \times 10^{-17} M(\text{gas}) \tag{13.24}$$

The rate of condensation of mass into supernovae and supermassive objects is

$$\frac{dM(SN + SM)}{dt} = 1.3 \times 10^{-17}\ M(gas)\ f\ (10) \qquad (13.25)$$

where f(10) is given by equation (13.23) with the appropriate M_1 which is itself a function of grain abundance - see equation (13.04). If we now ignore the differences between supermassive objects and supernovae (since the latter probably greatly predominate in grain production) then the rate of ejection of grains is

$$\frac{dM(\text{grains ejected})}{dt} \approx 1.3 \times 10^{-17}\ M(gas).f(10).F.ZSN. \qquad (13.26)$$

At the same time, grains are being lost with the gas condensing into stars, at a rate

$$\frac{dM(\text{grains lost})}{dt} = -\frac{dM(stars)}{dt}\left(\frac{M(grains)}{M(gas)}\right)$$

$$= 1.3 \times 10^{-17}\ M(gas)\left(\frac{M(grains)}{M(gas)}\right) \qquad (13.27)$$

The rate of growth of the mass fraction of grains in the interstellar gas is

$$\frac{d}{dt}\left(\frac{M(grains)}{M(gas)}\right) = \frac{\frac{d}{dt}M(grains)}{M(gas)} - \frac{M(grains)}{M(gas)} \cdot \frac{\frac{d}{dt}M(gas)}{M(gas)} \qquad (13.28)$$

Substituting into equation (13.28) from equations (13.24), (13.26) and (13.27), and using equation (13.21), finally

$$\frac{d}{dt}\left(\frac{n_g}{n_H}\right) = 1.63 \times 10^{-26}\ f(10).F.ZSN \qquad (13.29)$$

This can be integrated to give the grain abundance as a function of time using the values given in (13.19) with M_1 from chapter 11. Note that when $M_1 = 0.1$, then by equation (13.23), $f(10) = 0.22$; as (n_g/n_H) increases from 10^{-13} to 10^{-12}, f(10) rises progressively to 0.34.

At $t = 10^{10}$ yr., equation (13.29) will evidently give a grain abundance

$$\frac{n_g}{n_H} \sim (1.63 \times 10^{-26})(0.3)(0.04)(0.06)(3 \times 10^{17}) \sim 3 \times 10^{-12}$$

The current value in the solar region of our Galaxy (see chapter 11) is $\sim 10^{-12}$.

The results of the calculations, taking account of the changing limit M_2 and of the differences between supermassive objects and supernovae, are shown on Fig. 13.02.

The grain abundance rises slowly in the initial 10^8 years during which time the formation and explosion of supermassive objects dominates the overall evolution of the galaxy. It rises more rapidly as the densities in the clouds rise to the level at which large numbers of supernovae are formed (Fig. 11.03 or Fig. 13.01) especially if the rate of contraction of clouds is delayed, through heating, so that clouds remain a longer time in that density range.

Starting with low grain abundances, that is with $(n_g/n_H)(t = o) << 10^{-12}$, the effect of the feedback loop (13.3) slightly narrows any initial differences, and at $t = 10^{10}$ yr the grain abundance is $(n_g/n_H) \approx 10^{-12}$ independently of the starting conditions. Only if the initial grain abundance is already ~ 10^{-12} is there a significant difference; then at 10^{10} yr the abundance is about three times higher.

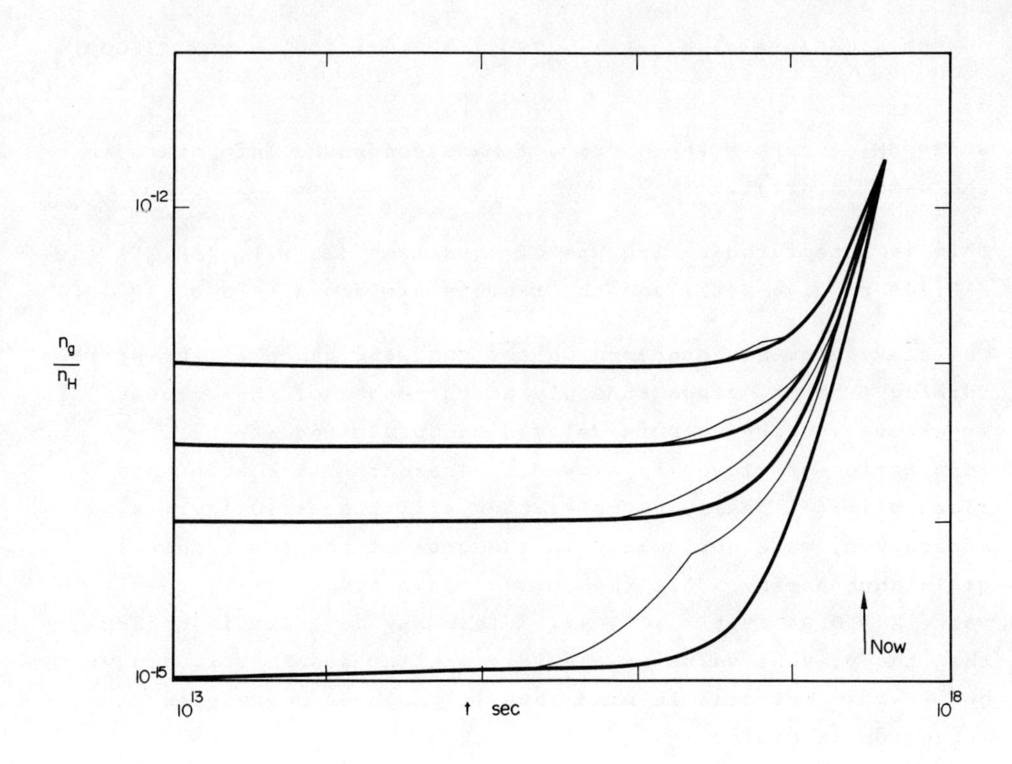

Fig. 13.02
The growth with time of interstellar grain abundance, for various initial values in the proto-galactic cloud. Heavy lines, cloud contraction at free fall rate, light lines at 10^{-2} free fall.

13.032 Abundance of Heavy Elements

The growth with time of the abundances of heavy elements is of course related to the grain abundance since the grains are formed from the heavy elements; in practice the detailed calculations of grain abundances reported in the previous section need to be carried out alongside calculations of heavy element abundances, since the latter determine the composition of the matter out of which stars form.

The enrichment of the gas by heavy elements in the time interval dt is

$$dZ = (dM(stars)/M(gas)) \, . \Big(ZSN.(f(30)-f(1000) + ZSM.f(1000)\Big) \qquad (13.30)$$

where dM(stars) is the amount of mass condensed into stars in the same interval.

This is integrated making use of equations (13.19), (13.23) and (13.24) to give Z(t), and the results are shown on Fig. 13.03.

The heavy element abundance in the gas, and hence in the stars forming from it, rises suddenly at the epoch of the outburst of supernovae, which is both delayed and prolonged if the contraction of clouds is slowed. Thereafter the abundance rises almost linearly, accelerating a little as 10^{10} yr is approached, more noticeably in the case of the lower initial grain abundances. All the curves converge, however, to give a value Z = 0.04 at t = 10^{10} yr. This may be a little higher than the present value in the interstellar gas in our Galaxy, but it must be borne in mind that it includes heavy elements locked up in grains.

At the time of formation of the Sun, that is t = 5 x 10^9 yr, the calculated abundance is Z = 0.01, in good agreement with the observed value.

It will be of interest subsequently to determine the abundance of heavy elements in stars of various masses and ages, and to find how many of them remain to be observed at 10^{10} yr.

13.033 Helium Abundance

The observational data relating to the abundance of helium in old stars is weak and the subject of much dispute. Although it is often argued in support of theories of a primeval origin

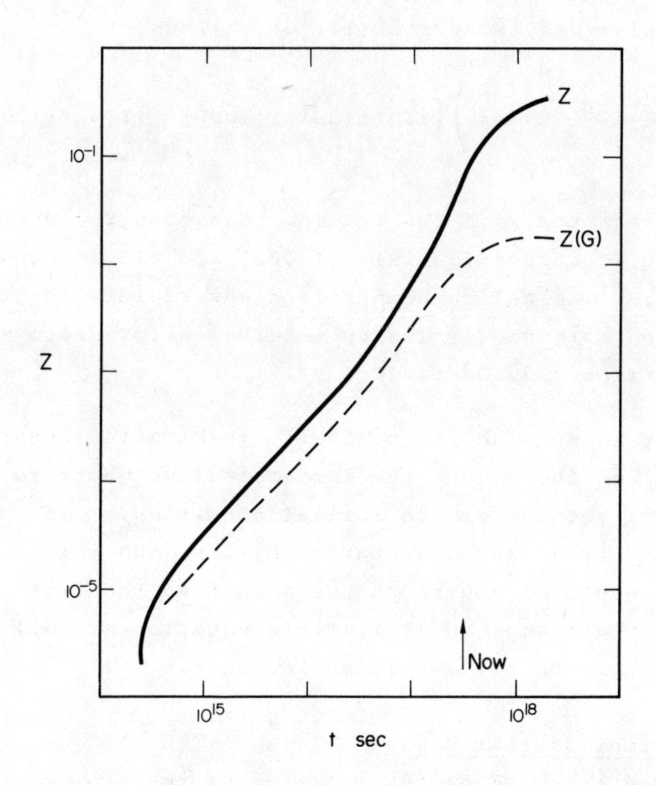

Fig. 13.03
The growth with time of heavy element abundance in the interstellar gas, Z. The average heavy element abundance in surviving dwarf G stars, Z(G) is shown by the broken line.

of most of the helium that the abundance in old stars is as high as in recently formed ones, the methods of determination are indirect. The only direct method, to which objections of interpretation can still be raised, is the analysis of matter ejected by the Sun which gives a helium abundance half the present value in a star formed half-way to the present epoch. A linear rise of helium abundance with time is indicated. These data are gathered together and discussed in chapter 7.

In the present model galaxy computations, the formation and calculation of the helium abundance is exactly analogous to that for heavy elements, equation (13.30). Thus,

$$\frac{dY}{dt} = \left(\frac{dM(stars)}{dt} \Big/ M(gas)\right)\left(YSN(f(30)-f(1000) \quad +YSM.f(1000)\right)(1-Y) \tag{13.31}$$

This is integrated with the initial condition Y = 0 at t = 0, again using equations (13.19), (13.23) and (13.24) to give Y(t), Fig. 13.04. The helium abundance rises rapidly for the first 10^7 yr, then more slowly for an interval before recovering a rapid rise from ~ 3.10^8 to 10^{10} yr.

This result is not inconsistent with the limited observational data described above, but the latter is inadequate to confirm or deny the correctness of the initial condition. If much of the helium production precedes star formation, the initial value on Fig. 13.04 would of course change more than the final value on account of the change in (1-Y)(t) in equation (13.31), so that the rate of growth of Y would be lessened.

13.034 Minimum Stellar Mass

A detailed discussion of the relation of the minimum proto-stellar mass M_1 to grain abundance has been given in chapter 11. The dependence of grain abundance on time (Fig. 13.02) transforms that relation to one with time, as shown on Fig. 13.05.

The upper mass limit M_2 is similarly related to grain abundance, but in practice statistical limitations set lower limits as described earlier in this chapter. These depend at first on the initial grain abundance and the total mass of the galaxy and are illustrated on Fig. 13.05 for $(n_g/n_H)(t = o) = 10^{-15}$ and $M(galaxy) = 10^{11}$ M⊙.

Fig. 13.05 shows the initial formation of supermassive objects in the first 10^7 yr, the outburst of supernovae, and the subsequent formation of stars of very low mass. It should be borne in mind that the frequency distribution of stars with masses

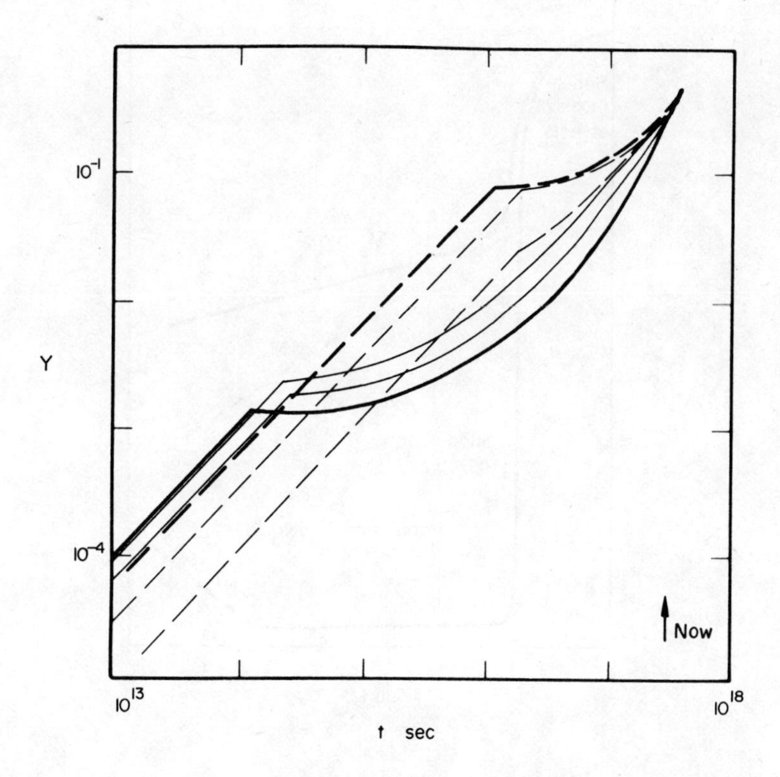

Fig. 13.04
The growth with time of helium abundance in the interstellar gas. The heavy lines are for an initial grain abundance of 10^{-15}, the light lines for 3×10^{-15} and 10^{-13}; continuous lines are for cloud contraction at free fall rate,; broken lines 10^{-2} of free fall. The results show that the rate of growth of helium abundance depends critically on the rate of contraction of clouds, but is insensitive to the initial grain abundance.

between the limits on Fig. 13.05 is that shown on Fig. 11.01, so that most of the stars have masses close to M_1 and the average mass $\overline{M} = 4M_1$.

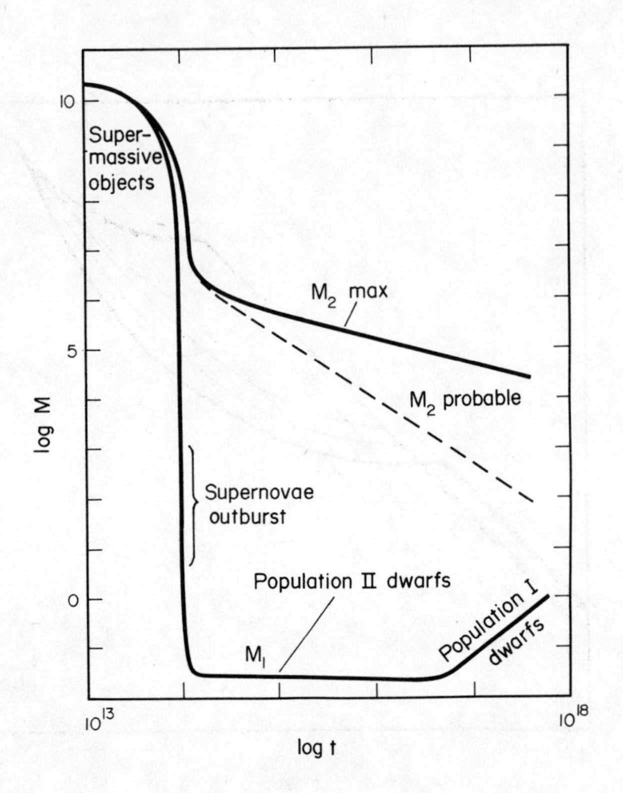

Fig. 13.05
The dependence on time of the limits to the masses of objects formed when cloud fragmentation results from the formation of hydrogen molecules on grain surfaces. The calculations show that supermassive objects form during the first 10^6 yr of evolution, followed by an outburst of supernovae and then the formation in turn of Population II dwarfs and Population I dwarfs. The range of masses of objects being formed at any one time is greatest at the beginning of the formation of Population II dwarfs, extending from about 0.03 M⊙ to 10^6 M⊙, and narrows progressively thereafter. It should cover a range from about 0.2 M⊙ to 100 M⊙ at the present time in our Galaxy.

13.035 Mass in Stars and Gas

If star formation results from the formation of H_2 molecules in the interstellar gas as proposed, then for given physical conditions it is to be expected that the rate of condensation of mass into stars will be directly proportional to the amount of interstellar gas available, as represented by equation (13.01)

$$\frac{d(M(stars))}{dt} = \frac{-d(M(gas))}{dt} = KM(gas) \qquad (13.32)$$

The observational data covering a wide range of values of M(gas) (chapter 6) indicates that this relationship is of the correct form, with $K = 1.3 \times 10^{-17}\ s^{-1}$ in the local group of galaxies. With this value for K, M(gas) declines and M(stars) increases with time as on Fig. 13.06. Account should be taken of the amount of mass ejected back into the interstellar gas by exploding stars, given by

$$\frac{dM(eject)}{dt} = KM(gas)\ f\ (M(limit)) \qquad (13.33)$$

where f(M(limit)), defined by equation (13.23) is the fraction of mass condensed into stars with masses above the limit for stability. This may be the white dwarf limit at about 1.4 $M_\odot$ but with $M_1 \sim 0.025$ this is a correction factor of at most 1.26 and is not substantial in comparison with the uncertainty in K and with the large effects of changes in K considered later.

Since

$$M(stars) = N\bar{M} = 4NM_1 \qquad (13.34)$$

the total numbers of stars formed can be found as a function of time from the data shown on Figs. 13.05 and 13.06, giving the result shown on Fig. 13.07. This shows that very large numbers of low mass dwarf stars are formed following the supermassive objects and the supernovae, during the time interval from $\sim 10^8$ yr to $\sim 3 \times 10^9$ yr.

13.036 Star Numbers and Heavy Element Abundances

The abundance of heavy elements as a function of time, Z(t), was calculated in section 13.032. Attempts to determine this

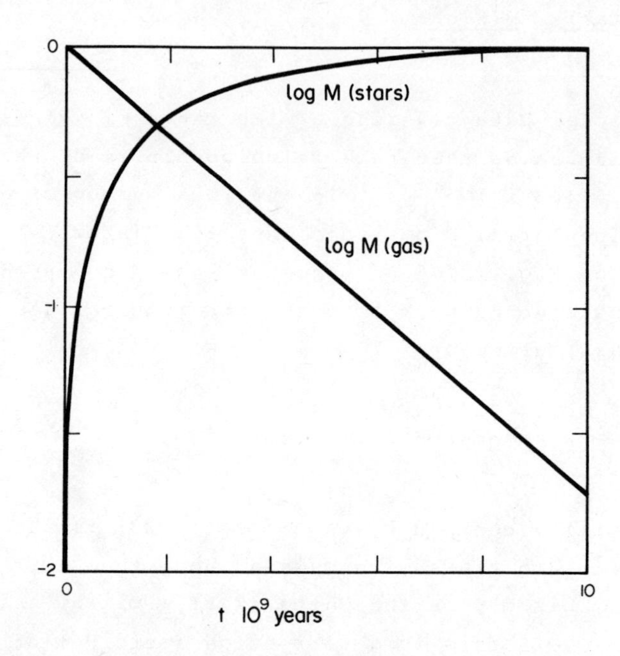

Fig. 13.06
The fraction of the proto-galactic cloud mass condensed into stars, and remaining in gas clouds, as functions of time.

relationship for stars in the galaxy by observation are handicapped by uncertainty in age determinations and selection effects and there are many anomalies and uncertainties in the results (chapter 7). Another approach is to calculate, for comparison with observation, the numbers of stars formed with heavy element abundances less than given values. This can be done using the data shown on Figs. 13.03 and 13.07 to give $N(\leqslant Z)$ at a given epoch, Fig. 13.08. This gives the median value of Z at $t = 10^{10}$ yr to be ~ 0.1%. About half the stars have abundances a tenth or less of that in the Sun; about 1 star in 14 has an abundance less than a hundredth of the Sun

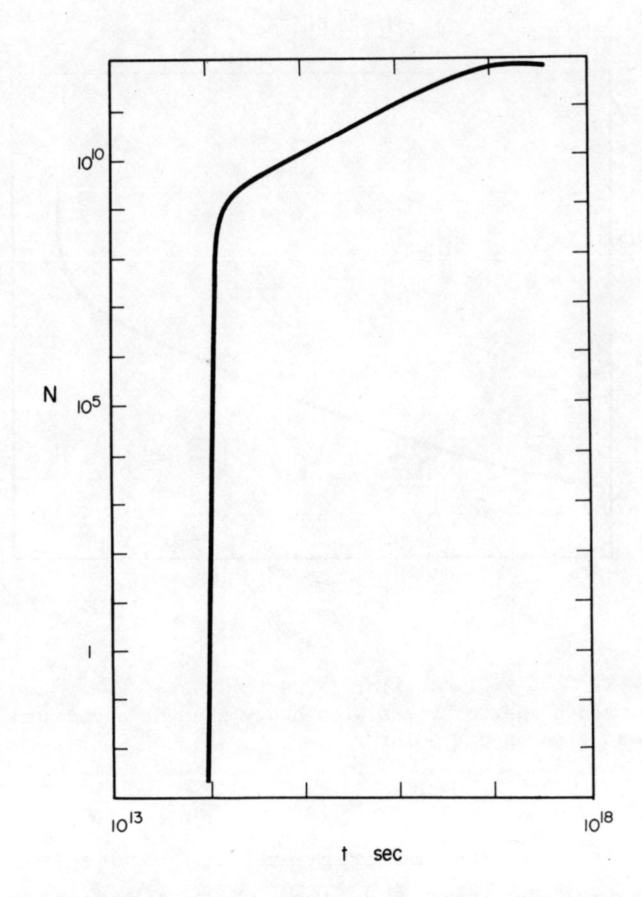

Fig. 13.07
The total numbers of stars formed, against time.

(extreme Population II) and 1 star in 300 has less than a thousandth of the solar heavy element abundance. Note that this refers to stars <u>formed</u>. However, not all stars formed

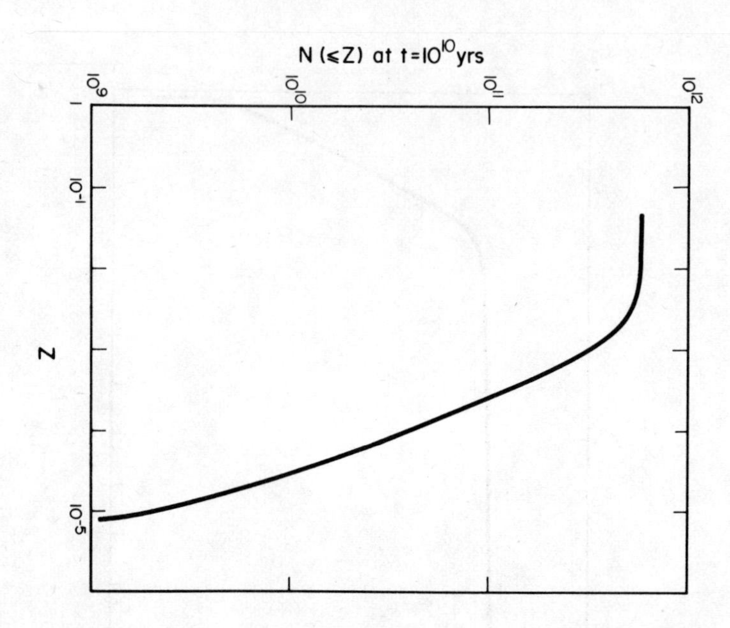

Fig. 13.08
The predicted number of stars with heavy element abundances $\leqslant$ Z at the present time in our Galaxy.

survive to 10^{10} yr. The older stars with masses $\geqslant 1\ M_\odot$ will have evolved to degeneracy and since these older stars have the lowest heavy element abundances the median Z for surviving stars will be rather larger. These results are in reasonably good agreement with the observational data reported in section 7.082.

Furthermore, if we consider comparison of the calculations with results based on analyses of the spectra of integrated light of galaxies, a selection effect occurs in the spectrophotometry by reason of the preponderant contribution by particular spectral types of stars to particular wavelength regions. For instance, O and B stars radiate primarily in the ultraviolet and blue, M stars in the near infrared, and G stars in the yellow. It is the latter that most observations select and a similar selection may be made in the calculations. The average heavy element abundance in surviving dwarf G stars at time t_1 is given by

$$\overline{Z}(G,t_1) = \int_o^{t_1} (dN(G)/dt) Z(t)\, dt \Bigg/ \int_o^{t_1} (dN(G)/dt)\, dt \qquad (13.35)$$

where Z(t) is given by the integration of equation (13.30) shown on Fig. 13.03, and from equation (13.02)

$$dN(G)/dt = M_1^{4/3} \left(M_L^{-4/3} - M_u^{-4/3} \right) dN/dt \qquad (13.36)$$

where M_L, M_u are the lower and upper limits to the masses of dwarf G stars. The integration of equation (13.35) is subject to the following conditions:

if $M_u \leqslant M_1$ or if $M_L \geqslant M(t_1,t)$ then $dN(G)/dt = o$;

$M_u > M_1 > M_L$ then M_L becomes M_1;

$M_u > M(t_1,t) > M_L$ then M_u becomes $M(t_1,t)$.

These conditions restrict the integration to stars that formed in the mass range and survived to t_1. The result is shown on Fig. 13.3. At the present time $t_1 \sim 10^{10}$ yr, Z = 4% and $\overline{Z(G)}$ = 1% in good agreement with observation. Since Z(t) = 1% at 5×10^9 yr, at the time that the Sun was formed, it follows that $\overline{Z(G)} \approx Z_\odot$.

13.037 Mass/Luminosity Ratios of Galaxies

The sum of the luminosities of surviving stars at any time gives the luminosity of the galaxy. Given the luminosities and lifetimes of stars of known mass, the sum can be calculated using the rate of star formation equation (13.24) and the frequency distribution of stellar masses equation (13.18). Thus

$$L(t_1) = \int_o^{t_1} L(t_1,t)\frac{dN}{dt}dt \tag{13.37}$$

where

$$L(t_1,t) = \int_{M_1(t)}^{M_2(t_1,t)} N(M)L(M,t_1,t)\,dM \Big/ \int_{M_1(t)}^{M_2(t)} N(M)\,dM \tag{13.38}$$

where $L(M, t_1,t)$ is the luminosity at t_1 of a star of mass M formed at t, $M_1(t)$ and $M_2(t)$ are the lower and upper limits to the masses of stars formed at t, and $M_2(t_1,t)$ is the mass of a star formed at t with lifetime $(t_1 - t)$. More massive stars than this will have shorter lifetimes and will not survive to t_1.

The dependence of stellar visual luminosity on mass is taken to be (Reddish, 1962a)

$$\left.\begin{aligned} L &= M^{3\cdot 85} \text{ for } M \leqslant 3.9\ M_\odot \\ L &= 15.2 \times M^{1\cdot 85} \text{ for } M > 3.9\ M_\odot \end{aligned}\right\} \tag{13.39}$$

where L, M are in solar units.

The lifetime of a star is approximated by

$$t_1 - t = 3 \times 10^{17}\ M/L \text{ s} \tag{13.40}$$

which represents the conversion of 10% of the stellar hydrogen to helium at constant luminosity. This is a reasonable approximation for middle and lower main sequence stars but it will underestimate the lifetime of massive stars; however, it turns out that most of the luminosity is contributed by stars of moderate mass.

Using the values for M_1 and M_2 derived earlier and shown on Fig. 13.05 the luminosity per unit mass of a galaxy can be calculated and is shown on Fig. 13.09.

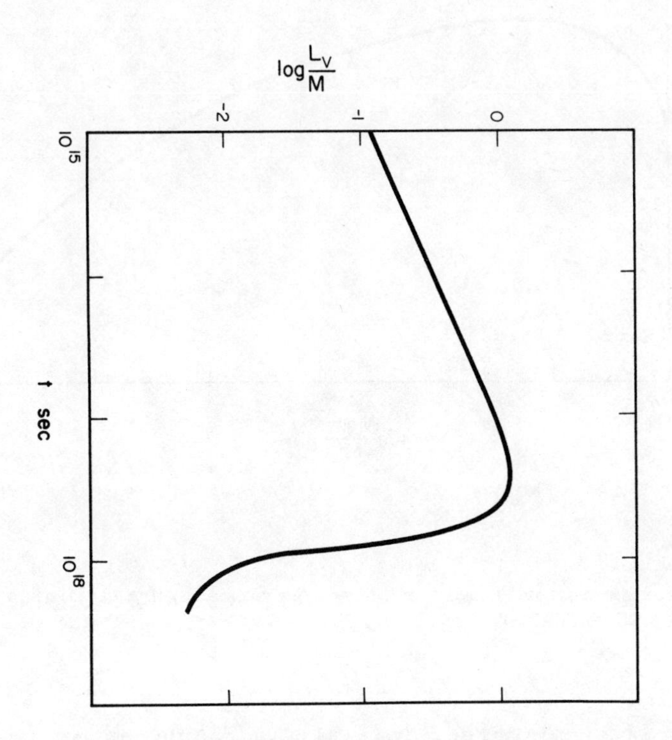

Fig. 13.09
The calculated luminosity per unit mass of a galaxy, against time. The rapid fall beginning at about 2 x 10^{10} yr results from the burning out of highly luminous stars and the absence of sufficient remaining gas to produce significant numbers of new ones.

Since the ages of galaxies are not known it is more useful for comparison with observation to show the relationship of luminosity to gas content using the data on Figs. 13.06 and 13.09.

The result given on Fig. 13.10 is not incompatible with observational data, but an explanation more probable than differences in age emerges in section 13.04.

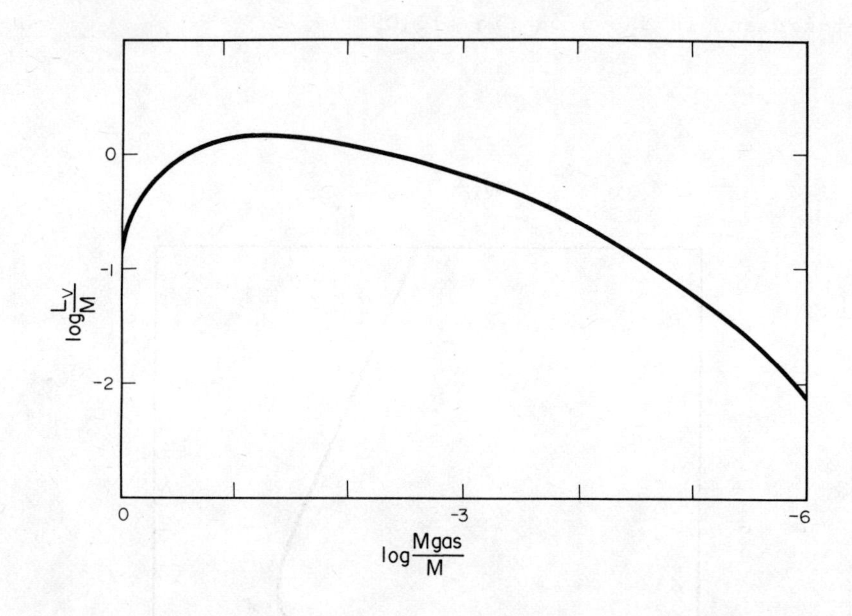

Fig. 13.10
The luminosity per unit mass against gas content, by combining Figures 13.06 and 13.09.

13.038 <u>Effects of Different Initial Grain Abundances</u>

The results have been presented mainly for model calculations with an initial grain abundance $(n_g/n_H)(t = o) = 10^{-15}$, although some of the Figures show the consequences of different initial values. It is apparent from these Figures that widely different initial values have little effect on the properties of galaxies 10^{10} yr later. A reduction in the initial grain abundance by three orders of magnitude reduces the grain and heavy element abundances at $t = 10^{10}$ yr by only a factor two. The effect on the helium abundance is never more than about a factor 2 and is negligible at 10^{10} yr. Although such a change in initial grain abundance causes the luminosity per unit mass of a galaxy to be lower by a factor ten at 3×10^7 yr, this has diminished to a factor 2 by 10^{10} yr. The value $(n_g/n_H)(t = o)$ is the only initial condition chosen arbitrarily, and the results are shown to be very insensitive to it.

13.040 Overshoot in the Rate of Star Formation

In the calculations already described the value used for the rate of condensation of mass into stars per unit mass of gas,

$$K = 1.3 \times 10^{-17}\ s^{-1}, \tag{13.41}$$

was derived directly from star counts and 21 cm HI measures in our Galaxy and in the local group of galaxies. We can make other estimates of it. From equation (13.01) we have

$$K = -2.30 \log_{10}(M(gas)/M_O)/t. \tag{13.42}$$

In our Galaxy, $M(gas) \approx 1.5\ M(HI) = 5 \times 10^9\ M_\odot$, and $M_O = 1.8 \times 10^{11}\ M_\odot$; putting $t = 10^{10}$ yr, equation (13.42) gives

$$K = 1.3 \times 10^{-17}\ s^{-1} \tag{13.43}$$

in agreement with the earlier value.

The average value of $M(gas)/M_O$ in other galaxies is 0.013. The data are given in Table 13.03, where the relative numbers are from Pskovskii (1961), the masses from Page (1965), and the gas fractions from Davies (1968a). The hydrogen contents given by Davies have been multiplied by 1.5 to allow for a helium content similar to that in our Galaxy. If we assume that all these galaxies have ages of 10^{10} yr the following results ensue. The mean value of K for other galaxies (all types) is, from equation (13.42),

$$\bar{K} = 1.44 \times 10^{-17}\ s^{-1}. \tag{13.44}$$

The mean gas content of irregular and spiral galaxies is $M(gas)/M_{O} = 0.040$, giving

$$\overline{K}(Irr + S) = 1.07 \times 10^{-17}\ s^{-1} \tag{13.45}$$

and the mean value for E galaxies

$$\overline{K}(E) > 2.8 \times 10^{-17}\ s^{-1} \tag{13.46}$$

approximately, assuming that the gas content of SO galaxies represents about the detectable limit.

TABLE 13.03

(1) Type of Galaxy	(2) Relative Number	(3) Average Mass	(4) Fractional Contribution to Mass Density	(5) $\frac{1.5 \times MHI}{M_O}$	(6) (4) x (5)
Irr+Sc	0.53	5	0.13	0.09	0.0117
Sb	0.13	10	0.06	0.015	0.0009
SO	0.13	30	0.13	0.0015	0.000195
E	0.21	70	0.68	<0.0015	<0.001
					0.013

It was pointed out earlier that the rate of formation of stars from molecular hydrogen may overshoot due to delay in the negative feedback in the loop (11.02); it is therefore of interest to recalculate the luminosities, gas contents, element and grain abundances in the model galaxies for different rates of star formation K. This has been done in the manner described earlier, where K had the single value $1.3 \times 10^{-17}\ s^{-1}$. The results of the new calculations based on the same data equations (13.19) are given in Table 13.04.

TABLE 13.04

Dependence of Results on K at $t = 10^{10}$ yr

K, 10^{-17} s^{-1}	log(M/L)	$\log(M_{gas}/M)$	log Z	log Y	$\log(n_g/n_H)$	$\log\overline{(Z(G))}$
0.3	0.23	-0.35	-2.61	-1.62	-14.20	-2.96
1.3	0.01	-1.52	-1.44	-0.52	-11.89	-2.08
2.3	0.00	-2.68	-0.86	-0.11	-10.64	-2.08
3.0	1.40	-3.50	-0.68	-0.06	-10.31	-2.07
4.0	1.43	-4.67	-0.68	-0.06	-10.31	-2.07
5.0	1.45	-5.83	-0.66	-0.05	-10.26	-2.08
6.0	1.45	-7.00	-0.66	-0.05	-10.27	-2.08

The remarkable result emerges that over a wide range of rates of star formation K, the mass/luminosity ratio takes up two discrete values,

$$\left.\begin{array}{l} M/L \approx 1 \quad \text{for } 0.3 \lesssim K \lesssim 2.5 \times 10^{-17}\ s^{-1} \\ M/L \approx 27 \text{ for } K \gtrsim 3 \times 10^{-17}\ s^{-1} \end{array}\right\} \qquad (13.47)$$

These are in good agreement with the observational data for irregular and spiral galaxies, and for E components of binary galaxies, respectively (Page loc. cit.). This result is a consequence of the exponential dependence of M(gas) on K. For $K < 2.5 \times 10^{-17}$ most of the luminosity is supplied by stars of moderate age and brightness. For $K > 3 \times 10^{-17}\ s^{-1}$ the amount of interstellar gas has declined so much that at 10^{10} yr there are relatively few stars of moderate age and most of the luminosity is supplied by the vast numbers of old dwarf stars.

A second notable feature of the results is the constancy in the mean heavy element abundance in surviving dwarf G stars $\overline{Z(G)}$ which has the value $\approx$ 1% for all $K \gtrsim 1.0 \times 10^{-17}\ s^{-1}$.

Although the heavy element abundance in the interstellar material at $t = 10^{10}$ yr, Z, reaches the high limiting value of 20% for large values of K, there is very little gas left by the time it approaches that, and a negligible number of stars are formed with such a large heavy element abundance.

The negative feedback loops still have the effect that most of the luminosity is produced by stars with approximately solar abundances. The constancy in the apparent heavy element abundances of galaxies, at about the solar value, is a feature of the observation data (Wood, 1966).

An upper limit to K if $\overline{Z(G)}$ is to remain $\approx$ 1% is set by the requirement that the time scale for condensation of the proto-galactic gas cloud into stars is much greater than the time scale for evolution of a single generation of supermassive stars and supernovae, which is ~ 10^6 yr. It follows from equation (13.42) that $K \not> 10^{-15}$ s^{-1}.

If $K \geqslant 3 \times 10^{-16}$ effectively the whole of the gas may condense into stars in 10^8 yr; this is less than the gravitational collapse time for the galaxy as a whole, and the stars will, therefore, occupy the volume of the 'proto-galactic' cloud instead of being collapsed to a disc, thus presenting the appearance of an elliptical galaxy; the smaller the value of K, the greater the time that elapses before effectively all the gas condenses into stars and the greater the degree of flattening which the cloud will have taken up.

Since the protostar globules of molecular hydrogen have much less internal pressure than their surroundings they will be compressed by the ambient gas and will therefore have a shorter contraction time than the galaxy as a whole - even if the galaxy as a whole is in free-fall collapse.

When $K \geqslant 3 \times 10^{-17}$ s^{-1} the value $\overline{Z(G)} \approx 1\%$ is reached in 5×10^9 yr and does not change significantly thereafter; when $K \geqslant 6 \times 10^{-17}$ s, M/L reaches 13 at $t = 5 \times 10^9$ yr, and then rises slowly through 27 at 10^{10} yr to 57 at 1.7×10^{10} yr. It is, therefore, possible that elliptical galaxies have a wide range of ages, some possibly less than 10^{10} yr and, therefore, younger than our Galaxy.

It is concluded that the properties of elliptical galaxies can result from overshoot in the rate of star formation by a factor of two or more, which may result from delay in the negative feedback in the loop grain cooling → molecule formation → star formation → grain heating. The results are summarized in Table 13.05 and Fig. 13.11.

13.05 Other Models and Analyses

Many calculations of galactic evolution have been made using empirical formulae for the frequency distribution of stellar

TABLE 13.05

Properties of model galaxies at age $t = 10^{10}$ yr

$\overline{K} = \overline{(dM(stars)/dt)/M(gas)} = 1.3 \times 10^{-17}\ s^{-1}$

	$0.23\overline{K} \leqslant K \leqslant 1.8\overline{K}$	$100\overline{K} \geqslant K \geqslant 2.3\overline{K}$
M/L	1	27
M(gas)/M	3%	<0.03%
Z(interstellar)	4%	20%
Z(at 5×10^9 yr)	1%	
$\overline{Z(G)}$	1%	1%
n_g/n_H	10^{-12}	5×10^{-11}
	Irr and S galaxies	E galaxies

masses, the rate of star formation as a function of the mass or density of interstellar gas, and so on, with the object of finding which empirical formulae lead to results best fitting observation (see the references at the beginning of this chapter) Generally the conclusions lend support to the theory outlined earlier in this and the preceeding two chapters, but one study in particular is of interest because it considers in some detail the manner in which star formation and galactic evolution may differ with position inside a galaxy (Larson, 1974). Before describing this it will be useful to note a number of points arising from other papers.

13.051 Grain Temperature

The temperature which grains may be expected to take up on average in giant elliptical galaxies, due to absorption of star-light and re-emission at infrared wavelengths, has been calculated by Hillel (1973). He concludes that it will be

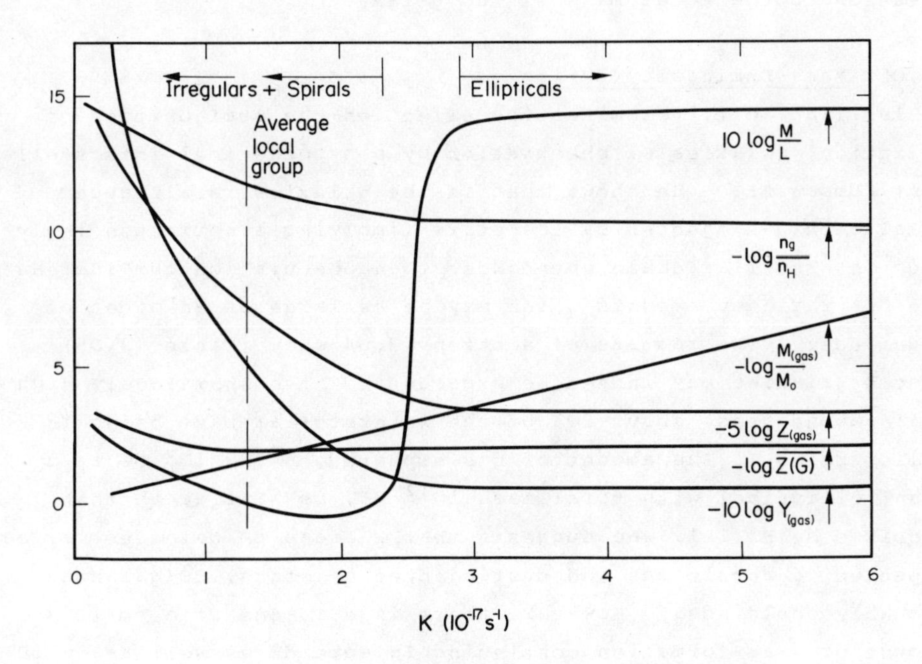

Fig. 13.11
The evolution of properties of a galaxy of stars, in which star formation results from cloud fragmentations induced by the formation of molecules on grain surfaces. The curves give the mass/luminosity ratio (M/L) and gas content (M(gas)/M) for the galaxy as a whole, the interstellar abundances of grains (n_g/n_H), heavy elements (Z) and helium (Y), and the average heavy element abundances in surviving dwarf G stars, all as functions of the fractional rate of condensation of gas into stars. The curves show that the properties of different types of galaxies are well accounted for by differences in this rate, which can be caused by very small differences in the effective temperature of the ambient radiation field (see Fig. 12.01).

$20 \leqslant T_g \leqslant 40K$ and that the galaxies will be strong emitters of infrared radiation at about 100 μm wavelength. The lower end of this range of temperature is about equal to the critical temperature for the formation of H_2 molecules on interstellar grains (Lee, 1972, 1975) indicating that molecule and hence star formation will proceed if gas is present. The amount of gas to be expected in elliptical galaxies is, however, very small and the rate of formation of new stars would be insignificant in relation to the total mass of the galaxy.

13.052 Mass-Luminosity Ratios

Hillel (loc. cit.) examines the effect on the luminosities of elliptical galaxies of obscuration by a hypothetical interstellar dust component. He shows that if the galaxies retain super metal rich gas ejected by the stars (implying a correspondingly high interstellar grain abundance) then obscuration can increase the M/L ratio by amounts which may be as large as an order of magnitude. The results of section 13.04 show (Table 13.05) that E galaxies may indeed be expected to have anomalously high grain abundances, about 20% of the interstellar mass being in solid grains. The amount of gas expected, $< 3 \times 10^8$ M⊙ in a giant elliptical with total mass 10^{12} M⊙, is similar to that required by Hillel, who suggests that whereas gE galaxies may be expected to retain gas and dust ejected by stars, dE galaxies probably would not. However, there is evidence of a small amount of star formation continuing in some dE as well as in gE galaxies (van den Bergh, 1973). No account was taken of possible obscuration by dust within galaxies when the values of M/L in Table 13.05 were calculated. They may need to be increased in the case of giant ellipticals.

13.053 Origin of Galaxy Types

It was shown in section 13.04 that the different integrated properties of E, S and I galaxies could probably be accounted for by a much higher initial rate of star formation in ellipticals and a rather lower one in irregulars, than in spirals. The same conclusion is reached by Matsuda (1970) who assumes that the rate of star formation depends on the second power of the gas density, and that the initial gas density in ellipticals is ten times higher, and in irregulars ten times lower, than in spirals. He therefore concludes that the different types of galaxies arise from differences in initial densities of the proto-galactic clouds. However, Bottinelli and Gouguenheim (1974) claim to have shown conclusively from a study of the properties of 131 galaxies that the type is not a consequence of initial density, but differences among galaxies of a given type may be. By contrast the different rates of star formation in section 13.04 arise from differences in grain temperature. As Fig. 12.01 shows, differences of only about 5% in the effective temperature of the ambient radiation field are sufficient to change the rates of star formation per unit mass of gas, K, by an order of magnitude, more that the difference between the K values of elliptical and spiral galaxies indicated by Table 13.05.

Consequently, although it appears now to be agreed that the different types of galaxies arise primarily from different rates of star formation per unit mass of gas, there is no agreement as to the origin of these different rates.

13.054 Chemical Evolution

The need for synthesis of elements in supermassive objects prior to the production of main sequence stars, if existing element abundances in the oldest stars formed are to be accounted for, has been emphasised by Silk and Siluk (1972). On the other hand, the relative abundances of different heavy elements are a strong indication that most of them were synthesised in the explosions of stars with masses $\geqslant 10\ M_\odot$, as assumed in the data equation (13.19) (Arnett, 1971).

Most of the empirical analyses referred to at the beginning of this chapter reach conclusions generally similar to those in section 13.032.

It appears to be fairly well established and widely agreed that in the early history of the galaxies the formation of numerous dwarf stars is preceded by the formation and explosion of relatively large numbers of massive stars and perhaps also by supermassive objects. As shown in section 13.01 this is to be expected on the basis of star formation resulting from the formation of H_2 molecules.

A somewhat different result in detail, although of the same general character, has been described by Ikeuchi et al, (1972). They point out that if mass loss by low mass stars is taken into account, the interstellar gas is diluted by matter which has not been enriched by nuclear synthesis in these stars and indeed, since many of the stars are old, is deficient in heavy elements in comparison with the interstellar gas at the epoch of ejection. Because of the long lifetimes of low mass stars this effect becomes more significant with time and the abundance of heavy elements in the interstellar gas and hence in the stars formed from it may reach a maximum and thereafter decline slowly. It is tempting to associate the epoch of the maximum, which may be as early as 10^9 yr, with the formation of super metal rich disc stars. This is an additional refinement to model galaxy calculations not considered earlier in this chapter.

Other calculations have also indicated that the chemical evolution of galaxies is insensitive to the time dependence of the rate of star formation, and that metal enriched star formation gives a better fit to observations than changing the initial mass function (Talbot and Arnett, 1973a, 1973b).

When the helium abundance is considered no unanimity of view emerges, in either observation or theory. Whether the helium abundance rose suddenly at an initial epoch or whether it has increased steadily with time is undecided (chapter 7), and whether a single cosmic event is required to produce it or many supermassive objects in galaxies may be sufficient is open to argument. It may be significant that, as shown in section 13.01, the formation of supermassive objects appears to be an inevitable consequence of the formation of H_2 molecules in clouds of proto-galactic mass.

13.055 <u>Evolution within Galaxies</u>

The studies described so far in this chapter have been concerned with total or average properties of galaxies. A recent study by Larson (1974) is of particular interest and importance because it calculates in considerable detail the variations with position within an evolving galaxy of such factors as the rate of star formation and the abundances of the elements.

The study relates to spherical galaxies and begins with a collapsing proto-galactic gas cloud, so that it has the same basis as the considerations in sections 13.01 to 13.03. It differs in that it goes on to consider separately the collapse rates of the gas and star distributions within which the internal motions are different; and it assumes that the rate of star formation depends on the rate of dissipation of the internal energy of motion of the gas, and hence on the gas density. The operation of some such effect was referred to in section 11.08, although only as an effect supplementary to the overriding control of the grain temperature on the rate of star formation. However, the high opacities required for grains to cool to the critical temperature for H_2 formation imply dense clouds, and the contraction of clouds to high densities implies the dissipation of internal energy of motion. Consequently the two studies may not differ in practice by as much as they may at first sight appear to differ in principle.

Since the gas density in Larson's model galaxy depends on position and time, and the rate of star formation depends on gas density, it follows that all of those features which stem from the formation and evolution of stars also depend on position and on time. The results are too extensive to reproduce in detail here, but some of the main points can be summarised. The gas density and hence the rate of star formation per unit mass of gas increase towards the centre of the galaxy, and this increase steepens with time. As a consequence the rate of enrichment of the gas by heavy elements increases towards the centre producing an inwardly increasing abundance of heavy elements in the stars and gas similar to that observed in galaxies. The abundances increase rather more sharply with time than in Fig. 13.03, but the assumed rate of star formation is higher and Larson's results are more appropriate to the models with higher K values in section 13.04 which reproduce the features of elliptical galaxies.

13.06 Conclusion

Many empirical analyses of the known data concerning our own and other galaxies (see references at the beginning of this chapter and section 13.05) have led to several main conclusions which are common to most or all of them. These are that in order to account for the observed numbers of stars of various masses and the abundances of the elements in them, the evolution of a galaxy goes through three stages (van den Bergh, 1973)

(a) The formation of very massive stars during the collapse of the proto-galaxy. These synthesise heavy elements and eject them into the interstellar material in a short period of time.

(b) The immediately subsequent formation of very large numbers of stars of low mass.

(c) The prolonged subsequent formation of substantial numbers of stars of intermediate mass.

These stages are prominent features of the model galaxies calculated in sections 13.01 to 13.03 of this chapter. Furthermore, the calculations have in addition produced numerical values for the abundances of interstellar grains, helium and heavy elements, and the differing properties of spiral and elliptical galaxies, which accord with observation. It is therefore of some interest to note that the models are based on a theory of star formation which in turn is based wholly on the physics of interstellar grains and molecule formation, a theory which predicts a stellar mass function in agreement with that observed. It contains no free or arbitrary parameters. Given a cloud of hydrogen with a low initial abundance of grains, and given the rate of cosmic expansion, it has proved possible using only the data of laboratory physics to construct a theory of star formation and to calculate the evolution of galaxies of stars thus formed. Indeed, considering the physics of molecule formation, the theory has an appearance not so much of possibility as of inevitability. The predictions of the theory and the results of the calculations are found to be in good agreement with the observed properties of stars and of galaxies of stars. It does not, of course, follow that the theory and the models are necessarily correct but it would be reasonable to suggest that they are perhaps of the right kind; it may be said of them that they have moved the study of the processes of star formation from the realms of empiricism and statistics to that of physics; and that may be counted progress.

REFERENCES

Ackermann, G., 1970. Mem. R. Soc. Sci. Liege, 19, 307.

Adams, W.S., 1949. Astrophys. J., 109, 354.

Aizenman, M.L., Demarque, P. & Miller, R.H., 1969. Astrophys. J., 155, 973.

Allen, C.W., 1964. Astrophysical Quantities, 2nd edn, Athlone Press, London.

Allen, D.A., 1972. Astrophys. J., 172, L55.

Alloin, D., 1970. Astr. Astrophys., 9, 45.

Altenhoff, W., 1968. Interstellar Ionized Hydrogen, p.519, ed Y. Terzian, Benjamin, New York.

Ambartsumyan, V.A., 1949. Astr. Zu., 26, 3.

Ambartsumyan, V.A., 1964. J. Br. astr. Ass., 75, 60.

Appenzeller, I., 1970. Astr. Astrophys., 5, 355.

Appenzeller, I. & Tscharnuter, W., 1974. Astr. Astrophys., 30, 423.

Arnett, W.D., 1971. Astrophys. J., 166, 153.

Arp, H.C., 1964. Astrophys. J., 139, 1045.

Arp, H.C., 1969a. Sky Telesc., 38, 385.

Arp, H.C., 1969b. Astr. Astrophys., 3, 418.

Arp, H.C., Burbidge, G.R. & Jones, T.W., 1973. Publ. astr. Soc. Pacific, 85, 423.

Auluck, F.C. & Kothari, D.S., 1965. Z. Astrophys., 63, 9.

Aveni, A.F. & Hunter, J.H., 1967. Astr. J., 72, 1019.

Baade, W., 1963. Evolution of Stars and Galaxies, p.59, Harvard University Press, Cambridge, Mass.

Baade, W. & Arp, H.C., 1964. Astrophys. J., 139, 1027.

Baglin, A., Berruyer, N., Morel, P.J. & Pecker, J.C., 1973. Astrophys. J., Lett., 15, 9.

Balkowski, C., Bottinelli, L., Gouguenheim, L. & Heidmann, J., 1973. Astr. Astrophys., 23, 139.

Ball, J.A. & Staelin, D.H., 1968. Astrophys. J., 153, L41.

Barrett, A.H., Schwartz, P.R. & Waters, J.W., 1971. Astrophys. J., 168, L101.

Barry, D.C. & Cromwell, R.H., 1974. Astrophys. J., 187, 107.

Baudel, L., 1970. Astr. Astrophys., 8, 65.

Becker, W., 1963. Z. Astrophys., 57, 117.

Becker, W., 1964. Z. Astrophys., 58, 202.

Becklin, E.E. & Neugebauer, G., 1968. Astrophys. J., 151, 145.

Bergh, S. van den., 1964. Astrophys. J. Suppl., 9, 65.

Bergh, S. van den., 1966. Astr. J., 71, 219.

Bergh, S. van den., 1967. J. R. astr. Soc. Can., 61, 23.

Bergh, S. van den., 1969a. Nature, 221, 48.

Bergh, S. van den., 1969b. Astrophys. Lett., 4, 117.

Bergh, S. van den., 1972. Publ. astr. Soc. Pacific, 84, 306.

Bergh, S. van den., 1973. J. R. astr. Soc. Can., 66, 237.

Bergh, S. van den. & Sher, D., 1960. Publ. David Dunlap Obs., 2, 203.

Berruyer, N., 1974. Astr. Astrophys., 30, 403.

Bessell, M.S., 1972. Publ. astr. Soc. Pacific, 84, 489.

Bingham, R.G. & Shakeshaft, J.R., 1967. Mon. Not. R. astr. Soc., 136, 347.

Bird, J.F., 1964. Rev. Phys., 36, 717.

Bird, J.F., 1968. Nature, 217, 1239.

Blaauw, A., 1964. A. Rev. astr. Astrophys., 2, 213.

Blanco, V.M. & Williams, A.D., 1959. Astrophys. J., 130, 482.

Bok, B.J., 1948. Harvard Obs. Monograph, No. 7, p.53.

Bok, B.J., 1956. Astr. J., 61, 309.

Bok, B.J., Cordwell, C.S. & Cromwell, R.H., 1970. Steward Observatory preprint No. 26.

Bok, B.J. & McCarthy, C.C., 1974. Astr. J., 79, 42.

Bok, B.J. & Reilly, E.F., 1947. Astrophys. J., 105, 255.

Bond, H.E., 1970. Astrophys. J., 22, 117.

Booth, R.S., 1969. Nature, 224, 783.

Borgman, J., Koornneef, J. & Slingerland, J., 1970. Astr. Astrophys., 4, 248.

Bottinelli, L., Chamaraux, P., Gouguenheim, L. & Heidmann, J., 1973. Astr. Astrophys., 29, 217.

Bottinelli, L. & Gouguenheim, L., 1974. Astr. Astrophys., 33, 269.

Bottinelli, L., Gouguenheim, L., Heidmann, J. & Heidmann, N., 1968. Ann. Astrophys., 31, 205.

Bridle, A.H., 1969. Nature, 221, 649.

Brooks, J.W., Murray, J.D. & Radhakrishnan, V., 1971. Astrophys. J. Lett., 8, 121.

Brück, M.T., Ireland, J.G., Nandy, K. & Reddish, V.C., 1968. Nature, 218, 662.

Bunner, A.N., Coleman, P.L., Kraushaar, W.L. & McCammon, D., 1971. Astrophys. J., 167, L3.

Burke, B.F., 1968. Interstellar Ionized Hydrogen, p.541, ed Y. Terzian, Benjamin, New York.

Burke, B.F., Papa, D.C. Papadopolous, G.D., Schwartz, P.R., Knowles, S.H., Sullivan, W.T., Meeks, M.L. & Moran, J.M., 1970. Astrophys. J., 160, L63.

Callebaut, D.K., 1967. Mem. Soc. R. Sci. Liege, 15, 319.

Carozzi, N., Chamaraux, P. & Duflot-Augarde, R., 1974. Astr. Astrophys., 30, 21.

Carranza, G., Courtes, G., Georgelin, Y., Monnet, G. and Pourcelot, A., 1968. Ann. Astrophys., 31, 63.

Carruthers, G.R., 1967. Astrophys. J., 148, L141.

Carruthers, G.R., 1970a. Space Sci. Rev., 10, 459.

Carruthers, G.R., 1970b. Astrophys. J., 161, L81.

Castle, K.G. & Hills, J.G., 1974. Astr. Astrophys., 30, 455.

Chaffee, F.H., Carbon, D.F. & Strom, S.E., 1971. Astrophys. J., 166, 593.

Chaisson, E.J., 1974. Astr. J., 79, 555.

Cheung, A.C., Rank, D.M., Townes, C.H., Thornton, D.D. & Welch, W.J., 1969. Nature, 221, 626.

Churchwell, E., Felli, M. & Mezger, P.G., 1969. Astrophys. Lett., 4, 33.

Clark, B.G., 1965. Astrophys. J., 142, 1398.

Clark, F.O., & Johnson, D.R., 1974. Astrophys. J., 191, L87.

Clausius, R., 1870. Phil. Mag., 40, 122, 4th series.

Clegg, R.E.S. & Bell, R.A., 1973. Mon. Not. R. astr. Soc., 163, 13.

Cohen, M., 1971. Astrophys. J., 169, 543.

Colgate, S.A., 1968. Astrophys. J., 153, 335.

Conti, P.S. & Loonen, J.P., 1970. Astr. Astrophys., 8, 197.

Courtes, G. & Dubout-Crillon, R., 1971. Astr. Astrophys., 11, 468.

Courtes, G., Georgelin, Y.P., Georgelin, Y.M. & Monnet, G., 1969. Astrophys. Lett., 4, 129.

Courtes, G. Georgelin, Y.P., Georgelin, Y.M., Monnet, G. & Pourcelot, A., 1968. Interstellar Ionized Hydrogen, p.571, ed Y. Terzian, Benjamin, New York.

Crutcher, R.M., 1973. Astrophys. J., 185, 857.

Cudaback, D.D. & Heiles, C., 1969. Astrophys. J., 155, L21.

Davies, R.D., 1957. Mon. Not. R. astr. Soc., 117, 663.

Davies, R.D., 1968a. Observatory, 88, 196.

Davies, R.D., 1968b. Nature, 218, 435.

Davies, R.D., Booth, R.S. & Wilson, A.J., 1968. Nature, 220, 1207.

Davies, R.D. & Gottesman, S.T., 1970. Mon. Not. R. astr. Soc., 149, 237.

Davies, R.D. & Lewis, B.M., 1973. Mon. Not. R. astr. Soc., 165, 231.

Davidson, K. & Harwit, M., 1967. Astrophys. J., 148, 443.

Demarque, P. & Schlesinger, B.M., 1969. Astrophys. J., 155, 965.

Dieter, N.H., 1967. Astrophys. J., 150, 435.

Dieter, N.H., 1969. Publ. astr. Soc. Pacific, 81, 186.

Disney, M.J., McNally, D. & Wright, A.E., 1969. Mon. Not. R. astr. Soc., 146, 123.

Dixon, M.E., 1966. Mon. Not. R. astr. Soc., 131, 325.

Dixon, M.E., 1967. Astr. J., 72, 429.

Dixon, M.E., 1970. Mon. Not. R. astr. Soc., 150, 195.

Dodd, R.J., 1974. Publ. R. Obs. Edin., 9, 63.

Dorschner, J., 1971. Astr. Nachr., 293, 65.

Dyson, J.E., 1973. Astr. Astrophys., 27, 459.

Eggen, O.J., 1967. A. Rev. Astr. Astrophys., 5, 105.

Eggen, O.J., 1969. Astrophys. J., 156, 241.

Eggen, O.J., 1970. Astrophys. J., 161, 159.

Eggen, O.J. & Sandage, A., 1969. Astrophys. J., 158, 669.

Eliasson, B. & Groth, E.J., 1968. Astrophys. J., 154, L17.

Ellder, J., Rönnäng, B. & Winnberg, A., 1969. Nature, 222, 67.

Fairall, A.P., 1971. Mon. Not. R. astr. Soc., 153, 383.

Falgarone, E. & Lequeux, J., 1973. Astr. Astrophys., 25, 253.

Faulkner, D.J. & Aller, L.H., 1965. Mon. Not. R. astr. Soc., 130, 393.

Feldman, P.A., Rees, M.J. & Werner, M.W., 1969. Nature, 224, 752.

Felli, M., Tofani, G. & D'Addario, L.R., 1974. Astr. Astrophys., 31, 431.

Field, G.B., 1962. The Distribution and Motion of Interstellar Matter in Galaxies, p.183, ed L. Woltjer, Benjamin, New York.

Field, G.B., Goldsmith, D.W. & Habing, H.J., 1969. Astrophys. J., 155, L149.

Field, G.B. & Saslaw, W.C., 1965. Astrophys. J., 142, 568.

FitzGerald, M.P., 1968. Astr. J., 73, 983.

Ford, W.K. & Rubin, V.C., 1968. Interstellar Ionized Hydroge, p.641, ed Y. Terzian, Benjamin, New York.

Foukal, P., 1969. Astrophys. Space Sci., 5, 469.

Fowler, W.A. & Hoyle, F., 1963. R. Obs. Bull. No. 67, 301

Gahm, G.F., & Winnberg, A., 1971. Astr. Astrophys., 13, 489.

Garzoli, S.L. & Varsavsky, C.M., 1966. Astrophys. J., 145, 79.

Gatley, I., Becklin, E.E., Matthews, K., Neugebauer, G., Penston, M.V. & Scoville, N., 1974. Astrophys. J., 191, L121.

Geyer, E.H., 1966. Veröff. Landessternw. Heidelb., 18, 1.

Gosachinskii, I.V., 1966. Sov. Astr., 9, 714.

Goss, W.M., 1968. Astrophys. J. Suppl., 15, 131.

Gottesman, S.T., Davies, R.D. & Reddish, V.C., 1966. Mon. Not. R. astr. Soc., 133, 359.

Gottesman, S.T. & Davies, R.D., 1970. Mon. Not. R. astr. Soc., 149, 263.

Grahl, B.H., Hachenberg, O. & Mebold, U., 1968. Beitr. Radio Astr., 1, 3.

Grzedzielski, S., 1966. Mon. Not. R. astr. Soc., 134, 109.

Guelin, M. & Weliachew, L., 1969. Astr. Astrophys., 1, 10.

Guibert, J., 1974. Astr. Astrophys., 30, 353.

Habing, H.J., 1969. Bull. astr. Inst. Netherl., 20, 177.

Habing, H.J. & Goldsmith, D.W., 1971. Astrophys. J., 166, 525.

Harman, R.J. & Seton, M.J., 1966. Astrophys. J., 140, 824.

Hartwick, F.D.A., 1971. Astrophys. J., 163, 431.

Hayakawa, S., Nishimura, S. & Takayanagi, K., 1961. Publ. astr. Soc. Japan, 13, 184.

Hearn, A.G., 1970. Mon. Not. R. astr. Soc., 150, 227.

Hearnshaw, J.B., 1972. Mem. R. astr. Soc., 77, 55.

Hegyi, D. & Curott, D., 1970. Phys. Rev. Lett., 24, 415.

Heiles, C., 1967a. Astrophys. J. Suppl., 15, 97.

Heiles, C., 1967b. Astrophys. J., 148, 299.

Heiles, C., 1969. Astrophys. J., 156, 493.

Heiles, C., 1970. Astrophys. J., 160, 51.

Herbig, G.H., 1962. Adv. astr. Astrophys., 1, 47.

Herbig, G.H., 1968. Z. Astrophys., 68, 243.

Herbig, G.H., 1970. Mem. Soc. R. Sci. Liege, 19, 13.

Hesser, J.E. & Henry, R.C., 1971. Publ. astr. Soc. Pacific, 83, 173.

Higgs, L.A. & Ramana, K.V.V., 1968. Astrophys. J., 154, 73.

Hillel, A.J., 1973. Astrophys. Space Sci., 25, 413.

Hills, J.G., 1972. Astr. Astrophys., 17, 155.

Hindman, J.V., 1967. Austr. J. Phys., 20, 147.

Hjellming, R.M. & Churchwell, E., 1969. Astrophys. Lett., 4, 165.

Hjellming, R.M., Gordon, C.P. & Gordon, K.J., 1969. Astr. Astrophys., 2, 202.

Hoag, A.A., Johnson, H.L., Iriarte, B., Mitchell, R.I., Hallam, K.L. & Sharpless, S., 1961. Publ. U.S. Nav. Obs., 17, 349.

Hobbs, L.M., 1974. Astrophys. J., 191, 395.

Hodge, P.W., 1966. An Atlas and Catalogue of HII Regions in Galaxies, University of Washington.

Hodge, P.W., 1969a. Astrophys. J., 155, 417.

Hodge, P.W., 1969b. Astrophys. J., 156, 847.

Hoerner, S. von., 1968. Interstellar Ionized Hydrogen, p.101, ed Y. Terzian, Benjamin, New York.

Hoffleit, D., 1953. Ann. Harv. Coll. Obs., 119, No. 2, 37.

Hoffleit, D., 1956. Astrophys. J., 124, 61.

Hoffmann, W.F., Frederick, C.L. & Emery, R.J., 1971. Astrophys. J., 164, L23.

Hoffmann, W.F., Frederick, C.L. & Emery, R.J., 1971. Astrophys. J., 170, L89.

Horedt, G.P., 1972. Astrophys. Space Sci., 15, 129.

Hoyle, F., 1960. Mon. Not. R. astr. Soc., 120, 22.

Hoyle, F. & Fowler, W.A., 1960. Astrophys. J., 132, 565.

Hoyle, F. & Ireland, J.G., 1961. Mon. Not. R. astr. Soc., 122, 35.

Hoyle, F. & Wickramasinghe, N.C., 1962. Mon. Not. R. astr. Soc., 124, 417.

Huang, Su-Shu., 1961. Publ. astr. Soc. Pacific., 73, 30.

Hughes, M.P., Thompson, A.R. & Colvin, R.S., 1971. Astrophys. J. Suppl. 23, 323.

Hulst, H.C. van de, Muller, H.C. & Oort, J.H., 1954. Bull. astr. Inst. Netherl., 12, 117.

Hunter, C., 1963. Mon. Not. R. astr. Soc., 126, 299.

Hunter, J.H., 1972. Mon. Not. R. astr. Soc., 157, 17P.

Ikeuchi, S., Sato, H., Sato, T. & Takeda, H., 1972. Progress theor. Phys., 48, 1885.

Ilovaisky, S.A. & Ryter, C., 1971. Astrophys., 15, 224.

Ireland, J.G., 1961. Mon. Not. R. astr. Soc., 122, 461.

Ireland, J.G., 1968. Z. Astrophys., 68, 294.

Isserstedt, J. & Schmidt-Kaler, T., 1968. Veröff. Bochum, Nr. 1.

Jager, G. de, & Davies, R.D., 1971. Mon. Not. R. astr. Soc., 153, 9.

Jeans, J.H., 1928. Astr. Cosmogony., Cambridge University Press, London.

Jenkins, E.B., 1969. IAU Symp. 36, 281.

Jones, B.F., 1970. Astr. J., 75, 563.

Jong, T. de, 1972. Astr. Astrophys., 20, 263.

Kalnajs, A.J., 1970. IAU Symp. 38, 318.

Kalnajs, A.J., 1972. Astrophys. J. Lett., 11, 41.

Kamp, P. van de, 1963. Astr. J., 68, 515.

Kayser, S.E., 1966. Astr. J., 72, 134.

Kellerman, K.I. & Pauliny-Toth, I.I.K., 1971. Astrophys. J., 166, L17.

Kemp, L.W., 1974. Astr. Astrophys. Suppl., 16, No.1.

Kerr, F.J., 1969. A. Rev. Astr. Astrophys., 7, 39.

Kerr, F.J., Burke, B.F., Reifenstein, E.C., Wilson, T.L. & Mezger, P.G., 1968. Nature, 220, 1210.

Kerr, F.J. & Garzoli, S., 1968. Astrophys. J., 152, 51.

Kerr, F.J., Hindman, J.V. & Gum, C.S., 1959. Austr. J. Phys., 12, 270.

Kholopov, P.N., 1958. Sov. Astr., 2, 398.

Kholopov, P.N., 1960a. Sov. Astr., 3, 291.

Kholopov, P.N., 1960b. Sov. Astr., 3, 425.

Kholopov, P.N., 1969. Sov. Astr., 12, 625.

Kiang, T., 1966. Z. Astrophys., 64, 433.

Kikuchi, S., 1967. Publ. astr. Soc. Japan, 19, 501.

Kilkenny, D., 1969. Private Communication.

Kippenhahm, R., 1969. Astr. Astrophys., 3, 83.

Klare, G., 1967. Z. Astrophys., 67, 131.

Klare, G. & Szeidl, B., 1966. Veröff. Landessternw. Heidelb., 18, 9.

Knapp, G.R., 1974. Astr. J., 79, 527.

Knowles, S.H., Meyer, C.H., Sullivan, W.T. & Cheung, A.C., 1969. Science, 166, 221.

Kolesnik, I.G. & Nadezhin, D.K., 1974. Astr. Zu., 51, 382.

Kossacki, K., 1968. Acta astr., 18, 221.

Kruszewski, A., 1961. Acta. astr., 11, 199.

Kushwaha, R.S. & Kothari, D.S., 1960. Z. Astrophys., 51, 11.

Lambert, D.L., 1967. Nature, 215, 43.

Lambert, D.L., 1968. Mon. Not. R. astr. Soc., 138, 143.

Lambrecht, H. & Schmidt, K.H., 1957. Astr. Nachr., 284, 71.

Larson, R.B., 1972a. Mon. Not. R. astr. Soc., 156, 437.

Larson, R.B., 1972b. Mon. Not. R. astr. Soc., 157, 121.

Larson, R.B., 1973. Mon. Not. R. astr. Soc., 161, 133.

Larson, R.B., 1974. Mon. Not. R. astr. Soc., 166, 585.

Larson, R.B., & Starrfield, S., 1971. Astr. Astrophys., 13, 190.

Lauque, R., 1973. Astr. Astrophys., 23, 253.

Layzer, D., 1964. A. Rev. Astr. Astrophys., 2, 341.

Lee, T.J., 1972. Nature Phys. Sci., 237, 99.

Lee, T.J., 1975. Astrophys. Space Sci., 34, 123.

Lequeux, J., 1970. Astrophys. J., 159, 459.

Lewis, B.M., 1972. Astr. Astrophys., 16, 165.

Lewis, B.M. & Davies, R.D., 1973. Mon. Not. R. astr. Soc., 165, 213.

Lin, C.C., 1967. A. Rev. Astr. Astrophys., 5, 453.

Lin, C.C. & Shu, F.H., 1964. Astrophys. J., 140, 646.

Lin, C.C. & Yuan, C., 1969. Astrophys. J., 155, 721.

Lin, C.C., Yuan, C. & Shu, F.H., 1969. Astrophys. J., 155, 721.

Lindblad, P.O., 1967. Bull. astr. Inst. Netherl., 19, 34.

Lindoff, U., 1968. Ark. Astr., 5, No. 1.

Lippincott, S.L., 1955. Astr. J., 60, 379.

Litvak, M.M., 1969. Astrophys. J., 156, 471.

Lozinskaya, T.A. & Kardashev, N.S., 1964. Soobshch. gos. astr. Inst. im. P.K. shternerga., No. 131, 37.

Luyten, W.J., 1956. Publ. astr. Soc. Pacific, 68, 258.

Luyten, W.J., 1968. Mon. Not. R. astr. Soc., 139, 221.

Lynden-Bell, D. & Ostriker, J.P., 1967. Mon. Not. R. astr. Soc., 136, 293.

Lynds, B.T., 1962. Astrophys. J. Suppl., 7, 1.

Lyngå, G., 1968. Proc. astr. Soc. Aust., 1, 92.

MacConnell, D.J., 1968. Astrophys. J. Suppl., 16, 275.

Madore, B.F., Bergh, S. van den, & Rogstad, D.H., 1974. Astrophys J., 191, 317.

Manchester, R.N. & Gordon, M.A., 1971. Astrophys. J., 169, 507.

Marcus, A.H., 1968. NASA CR-95938.

Marenco, G., Schutte, A., Scoles, G. & Tommasini, F., 1972. Trans. 5th Int. Vacuum Congress and Int. Conf. on Solid Surfaces, p.824.

Margon, B., 1974. Nature, 249, 24.

Martin, A.H.M. & Downes, D., 1972. Astrophys. J. Lett., 11, 219.

Matthews, W.G., 1969. Astrophys. J., 157, 583.

Mathewson, D.S., Kruit, P.C. & Brouw, W.N., 1971. Bull. Amer. astr. Soc., 3, 369.

Matsuda, T., 1970. Progress theor. Phys., 43, 1491.

McCuskey, S.W., 1956. Astrophys. J., 123, 458.

McCuskey, S.W., 1966. Vistas Astr., 7, 141.

McGee, R.X., Gardner, F.F. & Robinson, B.J., 1967. Aust. J. Phys., 20, 407.

McGee, R.X., & Milton, J.A., 1964. Aust. J. Phys., 17, 128.

McGee, R.X. & Milton, J.A., 1966. Aust. J. Phys., 19, 341.

MacGillivray, H., 1975. Mon. Not. R. astr. Soc., 170, 241.

Meaburn, J., 1971. Astrophys. Space Sci., 13, 110.

Mendoza, E.E., 1968. Astrophys. J., 151, 977.

Menon, T.K., 1958. Astrophys. J., 127, 28.

Menon, T.K., 1967. Astrophys. J., 150, L167.

Meszaros, P., 1968. Astrophys. Space Sci., 2, 510.

Meszaros, P., 1974. Nature, 248, 35.

Mezger, P.G., 1969. 6th IAU-IUTAM Joint Symposium on Cosmological Gas Dynamics & Gas Dynamics of the Interstellar Medium, Crimea, (IAU Symp. No. 39., Dordrecht, Reidel, 1970).

Mezger, P.G., 1970. Joint Discussion on Interstellar Molecules in Highlights of Astronomy, Vol. 2, ed C. de Jager, Reidel, Dordrecht.

Mezger, P.G., Altenhoff, W., Schraml, J., Burke, B.F., Reifenstein, E.C., & Wilson, T.L., 1967. Astrophys. J., 150, L157.

Mezger, P.G. & Hoglund, B., 1967. Astrophys. J., 147, 490.

Mezger, P.G. & Robinson, B.J., 1968. Nature, 220, 1107.

Mezger, P.G., Schraml, J. & Terzian, Y., 1967. Astrophys. J., 150, 807.

Mezger, P.G., Wilson, T.L., Gardner, F.F. & Milne, D.K., 1970a, Astrophys. Lett., 5, 117.

Mezger, P.G., Wilson, T.L., Gardner, F.F. & Milne, D.K., 1970b, Astrophys. Lett., 6, 35.

Mikulasek, Z., 1969. Bull. astr. Inst. Csl., 20, 215.

Miley, G.K., Turner, B.E. & Balick, B., 1970. Astrophys. J., 160, L119.

Minn, Y.K. & Greenberg, J.M., 1973. Astr. Astrophys., 22, 13.

Mitler, H.E., 1970. Smithsonian Astrophys. Obs. Special Report No. 323.

Morgan, W.W., 1968. In Interstellar Ionized Hydrogen, p.565, ed Y. Terzian, Benjamin, New York.

Morris, M., Palmer, P., Turner, B.E. & Zuckerman, B., 1974. Astrophys. J., 191, 349.

Muller, C.A. & Westerhout, G., 1957. Bull. astr. Inst. Netherl., 13, 15.

Murdin, P., 1969. Astrophys. J., 157, 175.

Murdin, P. & Sharpless, S., 1968. Interstellar Ionized Hydrogen, p.249. ed Y. Terzian, Benjamin, New York.

Murray, C.A. & Sanduleak, N., 1972. Mon. Not. R. astr. Soc., 157, 273.

Nakano, T., 1966. Progress theory Phys., 36, 515.

Nakano, T., 1973. Publ. astr. Soc. Japan, 25, 91.

Nakano, T. & Tademaru, E., 1972. Astrophys. J., 173, 87.

Neckel, T.,1966. Z. Astrophys., 63, 221.

Neugebauer, G. & Leighton, R.B., 1968. Sci. Am., 219, 50.

Ney, E.P. & Allen, D.A., 1969. Astrophys. J., 155, L193.

Nissen, P.E., 1970a. Astr. Astrophys., 6, 138.

Nissen, P.E., 1970b. Astr. Astrophys., 8, 476.

O'Connell, R.W. & Kraft, R.P., 1972. Astrophys. J., 175, 335.

Okuda, H. & Wickramasinghe, N.C., 1970. Nature, 226, 134.

Oort, J.H., 1954. Bull. astr. Inst. Netherl., 12, 177.

Oort, J.H., 1958a. Stellar Populations, p.415, Vatican City.

Oort, J.H., 1958b. Structure and Evolution of the Universe, p.163-183, ed R. Stoops, Solvay Institute, Brussels.

Oort, J.H., 1965a. Trans. IAU, 12A, 789.

Oort,J.H., 1965b. Galactic Structure, p.495, ed A. Blaauw and M. Schmidt, Chicago University Press.

Oort,J. H., 1970. Astr. Astrophys., 7, 381.

Oort, J.H. & Spitzer, L., 1955. Bull. astr. Inst. Netherl., 12, 177.

Osterbrock, D.E., 1970. Q. Jl. R. astr. Soc., 11, 199.

Paczynski, B., 1964. Acta Astr., 14, 157.

Paczynski, B., 1967. Acta Astr., 17, 1.

Page, T.L., 1965. Smithsonian Astrophys. Obs. Spec. Rep. No. 195.

Pagel, B.E.J., 1970a. Q. Jl. R. astr. Soc., 11, 172.

Pagel, B.E.J., 1970b. Vistas Astr., 12, 313.

Pagel, B.E.J., 1974. Mon. Not. R. astr. Soc., 167, 413.

Palmer, P., Zuckerman, B., Buhl, D. & Snyder, L.E., 1969. Astrophys. J., 156, L147.

Peebles, P.J.E., 1969. Astrophys. J., 157, 1075.

Peimbert, M., 1968. Astrophys. J., 154, 33.

Peimbert, M., 1970. Publ. astr. Soc. Pacific, 82, 636.

Peimbert, M. & Costero, R., 1969. Bol. Obs. Ton. y Tac., 5, 3.

Peimbert, M. & Spinrad, H., 1970a. Astr. Astrophys., 7, 311.

Peimbert, M. & Spinrad, H., 1970b. Astrophys. J., 159, 809.

Penston, M.V., 1974. Mon. Not. R. astr. Soc., 166, 21P.

Penston, M.V. & Allen, D.A., 1971. Astrophys. J., 170, L33.

Penston, M.V., Munday, V. Ann., Stickland, D.J. & Penston, M.J., 1969. Mon. Not. R. astr. Soc., 142, 355.

Perry, J.F.W. & Helfer, H.L., 1972. Astrophys. J., 174, 341.

Peters, G., 1970. Astr. Astrophys., 4, 134.

Pikelner, S.B., 1968. Sov. Astr., 11, 737.

Pikelner, S.B., 1970. Astrophys. Space Sci., 7, 489.

Popper, D.M., Jorgensen, H.E., Morton, D.C. & Leckrone, D.S., 1970. Astrophys. J., 161, L57.

Powell, A.L.T., 1970. Mon. Not. R. astr. Soc., 148, 477.

Powell, A.L.T., 1972. Mon. Not. R. astr. Soc., 155, 483.

Pratt, N.M., 1968. Publ. R. Obs. Edin., 6, 39.

Pratt, N.M., 1975. Private Communication.

Pronik, I.I., 1963. Izv. Krym. Astrofiz. Obs., 30, 118.

Pronik, I.I., 1974. Sov. Astr., 17, 457.

Przybylski, A., 1972. Mon. Not. R. astr. Soc., 159, 155.

Pskovskii, Yu. P., 1961. Sov. Astr., 5, 387.

Radhakrishnan, V. & Goss, W.M., 1971. Astrophys. J. Suppl., 170, 617.

Rappaport, S., Bradt, H.V. & Mayer, W., 1969. Astrophys. J., 157, L21.

Reddish, V.C., 1957. Observatory, 77, 202.

Reddish, V.C., 1958. Observatory, 78, 198.

Reddish, V.C., 1960. Proc. R. Soc. Edin., A65, 207.

Reddish, V.C., 1961. Observatory, 81, 19.

Reddish, V.C., 1962a. Sci. Prog., 50, 235.

Reddish, V.C., 1962b. Observatory, 82, 14.

Reddish, V.C., 1962c. Z. Astrophys., 56, 194.

Reddish, V.C., 1965. Vistas Astr., 7, 173.

Reddish, V.C., 1966. Nature, 211, 395.

Reddish, V.C., 1967a. Nature, 213, 1107.

Reddish, V.C., 1967b. Evolution of the Galaxies, Oliver & Boyd, Edinburgh.

Reddish, V.C., 1967c. Mon. Not. R. astr. Soc., 135, 251.

Reddish, V.C., 1968a. Rep. Astron. R. Scotl. for the year ending 31 March 1968, p.3.

Reddish, V.C., 1968b. Observatory, 88, 139.

Reddish, V.C., 1968c. Q. Jl. R. astr. Soc., 9, 409.

Reddish, V.C., 1968d. Interstellar Ionized Hydrogen, p.87, ed Y. Terzian, Benjamin Inc., New York.

Reddish, V.C., 1969a. Mon. Not. R. astr. Soc., 143, 139.

Reddish, V.C., 1969b. Mon. Not. R. astr. Soc., 145, 357.

Reddish, V.C., 1970. Mem. Soc. R. Sci. Liege, 19, 67.

Reddish, V.C., 1974. The Physics of Stellar Interiors, Edinburgh University Press.

Reddish, V.C., 1975a. Mon. Not. R. astr. Soc., 170, 261.

Reddish, V.C., 1975b. Unpublished.

Reddish, V.C., Lawrence, L.C. & Pratt, N.M., 1966. Publ. R. Obs. Edin., 5, 111.

Reddish, V.C. & Wickramasinghe, N.C., 1969. Mon. Not. R. astr. Soc., 143, 189.

Rickard, J.J., 1968. Astrophys. J., 152, 1019.

Riegel, K.W. & Crutcher, R.M., 1972a. Astrophys. J., 172, L107.

Riegel, K.W. & Crutcher, R.M., 1972b. Astr. Astrophys., 18, 55.

Robbins, R.R., 1970. Astrophys. J., 160, 519.

Roberts, M.S., 1957. Publ. astr. Soc. Pacific, 69, 59.

Roberts, M.S., 1968. Interstellar Ionized Hydrogen, p.617, ed Y. Terzian, Benjamin, New York.

Roberts, M.S., 1969. Astr. J., 74, 859.

Roberts, W.W., 1969. Astrophys. J., 158, 123.

Robertson, J.W., 1973. Astrophys. J., 185, 817.

Robinson, B.J., Gardner, F.F., van Damme, K.J. & Bolton, J.G., 1964. Nature, 202, 989.

Rogers, A.E.E. & Barret, A.H., 1963. Astrophys. J., 151, 163.

Rohlfs, K., 1967. Z. Astrophys., 66, 225.

Roshkovsky, D.A., 1954. Dokl. Acad. Nauk., 98, No. 4, 553.

Routly, P.M. & Spitzer, L., 1952. Astrophys. J., 115, 227.

Rubin, R.H., 1968. Astrophys. J., 153, 761.

Salpeter, E.E., 1955. Astrophys. J., 121, 161.

Sampson, W.B. & Reddish, V.C., 1975.

Sandquist, A., 1970. Astr. J., 75, 135.

Sanduleak, N., 1968. Astr. J., 73, 246.

Sanduleak, N., 1969. Astrophys. J., 74, 47.

Saraph, H.E., Seaton, M.J. & Shemming, J., 1969. Phil. Trans. R. Soc. London, A264, 78.

Sato, T., 1974. Kyoto University Preprint.

Schatzmann, E., 1958. IAU Symp. 8, Rev. Mod. Phys., 30, 1012.

Scheffler, H., 1967. Z. Astrophys., 65, 60.

Schmidt, K.H., 1971. Astr. Nach., 293, 57.

Schmidt, M., 1957. Bull. astr. Inst. Netherl., 13, 247.

Schmidt, M., 1959. Astrophys. J., 129, 243.

Schmidt, T., 1968. Z. Astrophys., 68, 380.

Schraml, J. & Mezger, P.G., 1969. Astrophys. J., 156, 269.

Schwarz, J., McCray, R. & Stern, R.F., 1972. Astrophys. J., 175, 673.

Schwarzschild, M. & Harm, R., 1959. Astrophys. J., 129, 637.

Scoville, N.Z. & Solomon, P.M., 1972. University Minnesota Astrophys. Reprint No. 12.

Scoville, N.Z. & Solomon, P.M., 1974. Astrophys. J., 187, L63.

Searle, L., 1971. Astrophys. J., 168, 327.

Searle, L., Sargent, W.L.W. & Bagnuolo, W.G., 1973. Astrophys. J., 179, 427.

Seaton, M.J., 1960. Rep. Prog. Phys., 23, 313.

Seddon, H., 1969. Nature, 222, 757.

Seyfert, C.K. & Nassau, J.J., 1945. Astrophys. J., 101, 179.

Sharpless, S., 1952. Astrophys. J., 116, 251.

Shklovsky, I.S., 1966. Astr. Tsirk., No. 372.

Shklovsky, I.S., 1967. Astr. Tsirk., No. 424, 1.

Shklovsky, I.S., 1968. Sov. Bloc Res. Geophys. Astronomy & Space, No. 178, 5.

Shklovsky, I.S., 1969. Sov. Astr., 13, 1.

Shu, F.H., Milione, V., Gebel, W., Yuan, C., Goldsmith, D.W. & Roberts, W.W., 1972. Astrophys. J., 173, 557.

Shu, F.H., Stacknik, R.V. & Yost, J.C., 1971. Astrophys. J., 166, 465.

Shuter, W.L.H. & Verschuur, G.L., 1964. Mon. Not. R. astr. Soc., 127, 387.

Silk, J. & Siluk, R.S., 1972. Astrophys. J., 175, 1.

Sim, M.E., 1968a. Publ. R. Obs. Edin., 6, 123.

Sim, M.E., 1968b. Publ. R. Obs. Edin., 6, 181.

Sim, M.E., 1968c. Observatory, 88, 62.

Simkin, S.M., 1970. Astrophys. J., 159, 463.

Simoda, M. & Kimula, H., 1968. Astrophys. J., 151, 133.

Simon, M., Righini, G., Joyce, R.R. & Gezari, D.Y., 1973. Astrophys.J., 186, L127.

Simonson, S.C., 1970. Astr. Astrophys., 9, 163.

Smith, A.M., 1969. Astrophys. J., 156, 93.

Smith, F.G., 1968. Nature, 220, 891.

Smith, L.F., 1968. Mon. Not. R. astr. Soc., 141, 317.

Smith, M.G. & Weedman, D.W., 1970. Astrophys. J., 161, 33.

Smith, M.G. & Weedman, D.W., 1971. Astrophys. J., 169, 271.

Snyder, L.E., Buhl, D., Zuckerman, B. & Palmer, P., 1969. Phys. Rev. Lett., 22, 679.

Spinrad, H., 1970. Q. Jl. R. astr. Soc., 11, 188.

Spinrad, H., Greenstein, J.C., Taylor, B.J. & King, I.R., 1970. Astrophys. J., 162, 891.

Spinrad, H., Gunn, J.E., Taylor, B.J., McClure, R.D. & Young,J.W., 1971. Astrophys. J., 164. 11.

Spinrad, H. & Peimbert, M., 1970. Galaxies and the Universe, chap.2, ed A. Sandage and M. Sandage, Chicago University Press.

Spinrad, H. & Taylor, B.J., 1969. Astrophys. J., 157, 1279.

Spinrad, H. & Taylor, B.J., 1971. Astrophys. J. Supp., 22, 445.

Spitzer, L., Drake, J.F., Jenkins, E.B., Morton, D.C., Rogerson, J.B. & York, D.G., 1973. Astrophys. J., 181, L116.

Steffey, P.C., 1968. Publ. astr. Soc. Pacific, 80, 565.

Steigman, G., Kozlovsky, B.Z. & Rees, M.J., 1974. Astr. Astrophys., 30, 87.

Stockton, A. & Chapman, G., 1970. Publ. astr. Soc. Pacific, 82, 306.

Strand, K.A., 1956. Astr. J., 61, 319.

Strom, K.M., Strom, S.E., Breger, M., Brooke, A.L. & Yost, J., 1972. Astrophys. J., 173, L65.

Strom, S.E., Strom, K.M., Brooke, A.L., Bregman, J. & Yost, J., 1972. Astrophys. J., 171, 267.

Strom, K.M., Strom, S.E. & Yost, J., 1971. Astrophys. J., 165, 479.

Taff, L.G. & Savedoff, M.P., 1973. Mon. Not. R. astr. Soc., 164, 357.

Talbot, R.J. & Arnett, W.D., 1973a. Astrophys. J., 186, 51.

Talbot, R.J. & Arnett, W.D., 1973b. Astrophys. J., 186, 69.

Tammann, G.A. & Sandage, A., 1968. Astrophys. J., 151, 825.

Terzian, Y. & Balick, B., 1969. Astr. J., 74, 76.

Tinsley, B.M., 1968. Astrophys. J., 151, 547.

Tinsley, B.M., 1972. Astrophys. J., 178, 319.

Tinsley, B.M., 1973. Astrophys. J., 186, 35.

Toomre, A., 1964. Astrophys. J., 139, 1217.

Toomre, A., 1969. Astrophys. J., 158, 899.

Torres-Peimbert, S., 1971. Bol. Obs. Ton. y Tac., 6, 3.

Torres-Peimbert, S., Lazcano-Araiyo, A., & Peimbert, M., 1974. Astrophys. J., 191. 401.

Torres-Peimbert, S. & Spinrad,H. 1971. Bol. Obs. Ton. y Tac., 6, 15.

Turner, B.E., 1969a. Astr. J., 74, 985.

Turner, B.E., 1969b. Astrophys. J., 157, 103.

Turner, B.E., 1970. Astrophys. J. Lett., 6, 99.

Unsold, A.O.J., 1969. Science, 163, 1015.

Varsavsky, C.M., 1968. Astrophys. J., 153, 627.

Vaucouleurs, G. de., 1955. Astr. J., 60, 126.

Venugopal, V.R. & Shuter, W.L.H., 1969. Mon. Not. R. astr. Soc., 143, 27.

Verschuur, G.L., 1969. Nature, 223, 140.

Verschuur, G.L., 1971. Astrophys. J. Lett., 7, 217.

Vorontsov-Velyaminov, B.A., 1969. Sov. Astr., 13, 235.

Wagoner, R.V., Fowler, W.A. & Hoyle, F., 1967. Astrophys. J., 148, 3.

Walker, M.F., 1964. Astr. J., 69, 744.

Walker, M.F., Blanco, V.M. & Kunkel, W.E., 1969. Astr. J., 74, 44.

Walmsley, M. & Grewing, M., 1971. Astrophys. J. Lett., 9, 185.

Wanner, J.F., 1972. Mon. Not. R. astr. Soc., 155, 463.

Wares, G.W. & Aller, L.H., 1968. Publ. astr. Soc. Pacific, 80, 568.

Warner, B., 1961a. Observatory, 81, 230.

Warner, B., 1961b. Publ. astr. Soc. Pacific, 73, 439.

Weinberg, S., 1971. Astrophys. J., 168, 175.

Weizsacker, C.F. von, 1951. Astrophys. J., 114, 165.

Welch, G.A., 1970. Astrophys. J., 161, 821.

Welch, G.A. & Forrester, W.T., 1972. Astr. J., 77, 333.

Werner, M.W. & Harwit, M., 1968. Astrophys. J., 154, 881.

Werner, M.W. & Salpeter, E.E., 1970. Mem. R. Soc. Sci. Liege, 19, 113.

Werner, M.W., Silk, J. & Rees, M.J., 1970. Astrophys. J., 161, 965.

Westerhout, G., 1957. Bull. astr. Inst. Netherl., 13, 201.

Westerlund, B.E., 1961. Publ. astr. Soc. Pacific, 73, 51.

Westerlund, B.E., 1969. Astr. J., 74, 882.

Westerlund, B.E. & Smith, L.F., 1964. Mon. Not. R. astr. Soc., 128, 311.

Whiteoak, J.B. & Gardner, F.F., 1969. Proc. astr. Soc. Aust., 1, 282.

Wickramasinghe, N.C., 1968. Observatory, 88, 246.

Wickramasinghe, N.C., 1973a. Light Scattering Functions for Small Particles, Adam Hilger, London.

Wickramasinghe, N.C., 1973b. Astrophys. Space Sci., 20, 123.

Williams, J.A., 1966. Astr. J., 71, 615.

Williams, R.E., 1967, Astrophys. J., 147, 556.

Wilson, W.J., 1971. Astrophys. J., 166, L13.

Wilson, W.J. & Barrett, A.H., 1968. Science, 161, 778.

Wilson, W.J. & Barrett, A.H., 1970. Astrophys. Lett., 6, 231.

Wing, R.F., Spinrad, H. & Kuhi, L.V., 1967. Astrophys. J., 147, 117.

Wood, D.B., 1966. Astrophys. J., 145, 36.

Woolley, R.v.d.R., 1960. Mon. Not. R. astr. Soc., 120, 214.

Woolley, R.v.d.R. & Stewart, J.M., 1967. Mon. Not. R. astr. Soc., 136, 329.

Wynn-Williams, C.G., 1971. Mon. Not. R. astr. Soc., 151, 397.

Wynn-Williams, C.G. & Becklin, E.E., 1974. Publ. astr. Soc. Pacific, 86, 5.

Wynn-Williams, C.G., Becklin, E.E. & Neugebauer, G., 1972. Mon. Not. R. astr. Soc., 160, 1.

Yentis, D.J., Novick, R. & Bout, P.v.d., 1972. Astrophys. J., 177, 365.

Yoneyama, T., 1972. Publ. astr. Soc. Japan, 24, 87.

Zuckerman, B. & Palmer, P., 1968. Astrophys. J., 153, L145.

SUBJECT INDEX